THE AFTERLIVES OF THE
PSYCHIATRIC ASYLUM

Geographies of Health

Series Editors
Allison Williams, Associate Professor, School of Geography and
Earth Sciences, McMaster University, Canada
Susan Elliott, Professor, Department of Geography and Environmental
Management and School of Public Health and Health Systems,
University of Waterloo, Canada

There is growing interest in the geographies of health and a continued interest in what has more traditionally been labeled medical geography. The traditional focus of 'medical geography' on areas such as disease ecology, health service provision and disease mapping (all of which continue to reflect a mainly quantitative approach to inquiry) has evolved to a focus on a broader, theoretically informed epistemology of health geographies in an expanded international reach. As a result, we now find this subdiscipline characterized by a strongly theoretically-informed research agenda, embracing a range of methods (quantitative; qualitative and the integration of the two) of inquiry concerned with questions of: risk; representation and meaning; inequality and power; culture and difference, among others. Health mapping and modeling has simultaneously been strengthened by the technical advances made in multilevel modeling, advanced spatial analytic methods and GIS, while further engaging in questions related to health inequalities, population health and environmental degradation.

This series publishes superior quality research monographs and edited collections representing contemporary applications in the field; this encompasses original research as well as advances in methods, techniques and theories. The *Geographies of Health* series will capture the interest of a broad body of scholars, within the social sciences, the health sciences and beyond.

Also in the series

Geographies of Health and Development
Edited by Isaac Luginaah and Rachel Bezner Kerr

Soundscapes of Wellbeing
in Popular Music
Gavin J. Andrews, Paul Kingsbury and Robin Kearns

Mobilities and Health
Anthony C. Gatrell

The Afterlives of the Psychiatric Asylum

Recycling Concepts, Sites and Memories

GRAHAM MOON
University of Southampton, UK

ROBIN KEARNS
University of Auckland, New Zealand

ALUN JOSEPH
University of Guelph, Canada

Routledge
Taylor & Francis Group

LONDON AND NEW YORK

First published 2015 by Ashgate Publishing

2 Park Square, Milton Park, Abingdon, Oxfordshire OX14 4RN
52 Vanderbilt Avenue, New York, NY 10017

Routledge is an imprint of the Taylor & Francis Group, an informa business

First issued in paperback 2020

British Library Cataloguing in Publication Data
A catalogue record for this book is available from the British Library

The Library of Congress has cataloged the printed edition as follows:
Moon, Graham, 1956-
  The afterlives of the psychiatric asylum : recycling concepts, sites and memories / by Graham Moon, Robin Kearns and Alun Joseph.
     pages cm. -- (Ashgate's geographies of health series)
  Includes bibliographical references and index.
  ISBN 978-1-4094-4252-3 (hardback)
  1. Psychiatric hospitals--History. 2. Mentally ill--Care. I. Kearns, Robin A., 1959- II. Joseph, A. E. (Alun E.) III. Title.
  RC439.M74 2015
  362.2'1--dc23

                                                                    2014049974

ISBN: 978-1-4094-4252-3 (hbk)
ISBN: 978-0-367-66880-8 (pbk)

Contents

List of Figures

List of Tables

Acknowledgments

As authors our first acknowledgement must be to our partners, Liz, Pat and Gill, who have patiently accommodated our interest in former asylums even when they found it quite strange. We appreciate your forbearance and support. We also gratefully acknowledge the editorial assistance of Tara Coleman, without whom the final months of writing would have been even more extended. Mark Dover and Igor Drecki provided their graphic and cartographic skills for which we are very appreciative. Colleagues at successive meetings of the International Symposium in Medical Geography provided useful feedback on emerging ideas. In particular, we wish to acknowledge informed insights from Andrew Power and Ronan Foley. Additional welcome insights have been provided over the years by Chris Philo. All these inputs have been a source of encouragement as our project developed. Finally, we would also wish to acknowledge the initial inspiration from John Giggs, Michael Dear, Martin Taylor and Chris Smith that led us each in different ways to our interest in mental health care provision.

This monograph provides an update, development and synthesis of research that has been ongoing for over a decade. Chapter 3 draws particularly on Moon, G., Kearns, R., and Joseph, A. (2006). Selling the private asylum: Therapeutic landscapes and the (re)valorization of confinement in the era of community care. *Transactions of the Institute of British Geographers*, 31(2), 131–49, Chapter 5 on Kearns, R., Joseph, A.E. and Moon, G. (2010). Memorialisation and remembrance: On strategic forgetting and the metamorphosis of psychiatric asylums into sites for tertiary educational provision. *Social & Cultural Geography*, 11(8), 731–49, and Chapter 6 on Joseph, A., Kearns, R. and Moon, G. (2013). Re-imagining psychiatric asylum spaces through residential redevelopment: Strategic forgetting and selective remembrance. *Housing Studies*, 28(1), 135–53. In each case we have extended our arguments and added new material.

The majority of graphics and photographs are our own or specially commissioned. We additionally acknowledge the reproduction of the following:

Figure 3.1: © Copyright Stu's Images, licensed for reuse under a Creative Commons Attribution-Share Alike 3.0 Unported license. *Source*: http://commons. wikimedia.org/wiki/File:Grovelands_House,_The_Bourne,_Southgate,_N14.jpg.

Figure 4.6: © Copyright Andrew Smith, licensed for reuse under a Creative Commons Attribution-Share Alike 2.0 Generic Licence. *Source*: http://www. geograph.org.uk/photo/106921.

Chapter 1

Introduction

For the great majority of these establishments [psychiatric asylums] there is no appropriate future use, and I for my own part will resist any attempt to foist another purpose upon them unless it can be proved to me in each case that, such, or almost such, a building would have had to be erected in that, or some similar, place to serve the other purpose, if the mental hospital had never existed (Powell, 1961).

In his 1961 opening address to the National Association for Mental Health, the then UK Secretary of State for Health, Enoch Powell, gave voice to what was to become one of his more felicitous contributions to society. He spoke of the imminent programme to close the network of psychiatric asylums and move to community-based care for people with mental health problems. Powell's speech, characteristically convoluted, is generally remembered, and indeed named, for its reference to the water towers that in the UK almost invariably signalled the presence of asylums within the landscape. Less noted is the above quote, with its assertion that there is no appropriate future use for asylums. In this book, we draw upon 20 years of collective collaborative research to assess Powell's bleak prognostication – primarily for psychiatric asylum sites but also for the idea of asylum – and to develop an understanding of the interrelated processes, operating at local and higher scales, that have acted to shape the fate of this treatment modality and the sites of its delivery. In so doing, we also wish to develop an appreciation of the traces of this once-dominant form of care in contemporary physical and cultural landscapes and of consequent implications for the remembrance of the psychiatric asylum.

Our subject matter demands an early statement about language. Mental health care is a fraught area in this regard, with linguistic preferences shifting and changing over time and place and also with respect to meaning. The facilities that are our central concern have been variously retreats, asylums, hospitals and homes. They have served lunatics, the insane, mad people, people with mental health problems and the mentally ill. More pejorative terms pervade popular discourse. We are acutely aware of the labelling and stereotyping that follows from this changing and contested terminology. In this book we use the term 'psychiatric asylum', which we contend to be neutral but meaningful. It conveys the focus of the institutions (psychiatric) and their underlying philosophy (asylum). For preference, we refer to people with mental health problems as the users of these facilities.

During an era of closure spanning the decades since the 1960s, with temporal epicentres varying by country, the psychiatric asylum was, at once, characterised as outmoded and ill-suited to the needs of contemporary health care and stigmatised as a site of patient abuse. This made it all too easy to forget that at its inception more than 150 years earlier the psychiatric asylum was itself an innovation of considerable significance. In the early nineteenth century, the options for families who could no longer care, or arrange care, for people with mental health problems at home were limited. The psychiatric asylum represented a rejection of practises that had seen such people relegated to poor houses, private establishments or prisons, and sometimes exploited as spectacle or as a source of profit (Philo, 2004). The emergence of the idea of asylum epitomised a desire for order that diffused from Europe to the rest of the world. It manifested itself in a categorisation through which people with mental health problems were seen to be distinct and different from the likes of offenders, the indigent poor and other groups on the margins of 'normal' society (Foucault, 1967). In later years this ordering would be extended to differentiate further between mental health problems and learning disabilities and between the needs of children and adults.

Key characteristics of the idea of the psychiatric asylum included the notion of potential recovery in an ordered, secluded and generally rural environment away from the social stresses of an emerging industrial urbanised society. Coincidentally, this also catered to the sensitivities of families and society at large regarding mental illness by facilitating the removal of individuals whose behaviour was deemed in some way to be socially unacceptable. In addition to a calming, therapeutic landscape, recovery in asylum settings was also held to require opportunities for structured work. Early proponents of this approach, such as Tuke at the York (UK) Retreat, were instrumental in establishing the positive case for asylum. By the middle of the nineteenth century the case was widely and internationally accepted, being adopted by the state as a response to the growing mental health care needs of burgeoning urban populations that could neither be accommodated in the erstwhile private asylums nor afford their fees.

State involvement in psychiatric care involved an inevitable scaling up of the 'private retreat' in order to accommodate larger numbers and to do so at a lower cost. The institutions created were large and visually impressive, with leading architects of the time being employed in their design. They were built within extensively landscaped estates in locations removed, but not always isolated, from cities and other major centres of population (see Chapter 4). England provides some idea of the eventual extent of the network of state-provided psychiatric asylums and the scale of constituent institutions. By 1914 there were over 100 asylums. The great majority of these were 'county asylums', with most of the then 39 traditional counties and many of the 39 county boroughs having at least one facility. A small minority of asylums continued as private or 'subscription' facilities. London was served by some 16 asylums. Notable among these was the 'Epsom colony', a cluster of five facilities occupying a 405 hectare site to the south-west of London. Single site facilities were no less extensive. The former Colney Hatch Hospital featured monolithic Italianate buildings in 13 hectares on the outskirts of London. Sites of 80–160 hectares were not uncommon. Asylum buildings were intended to impress and to stand out in their park-like grounds, and few failed in this regard. The upkeep of buildings and grounds and the care of patients provided considerable employment for those in nearby towns and cities and in the case of the more isolated asylums gave rise to single-purpose communities dedicated to the housing of asylum workers (Philo, 2004).

Across the world, the psychiatric asylum, made real in bricks and mortar, epitomised the delivery of mental health care well into the second half of the twentieth century, with many institutions surviving to celebrate their centenaries. It is these facilities that are the key focus of this book. The seeds of their fall from favour were evident early. Though built with the best of intentions to meet mental health needs, population growth was such that demand outstripped capacity within a few decades of their construction. Moreover, there was a realisation that for all its positive aspects, asylum did not necessarily result in cure for a significant subset of patients, with consequences for the throughput of resident numbers. Overcrowding emerged as a serious problem and was marked by the middle of the twentieth century. Yet the idea of asylum persisted – seclusion and separation remained a hallmark of mental health care. Although the monolithic asylum buildings in their isolated (and isolating) estates were supplemented by 'villa developments' intended to humanise the scale of provision but retain its key elements, the asylum nonetheless remained a key presence in the physical and cultural landscape into the latter half of the twentieth century.

While overcrowding and emergent evidence for ineffectiveness may have presaged the demise of the asylum, their well-documented final fall from favour resulted from further factors. The fine buildings of the nineteenth century were showing their age following years of heavy use by larger populations than were ever intended; repair costs were escalating. At the same time, earlier evidence for ineffectiveness was put into stark relief by the emerging availability of alternative treatments, notably those employing new drugs that enabled people to be treated outside institutional settings. Moreover, the institutional nature of the asylum was increasingly recognised as problematic. While isolation and seclusion may have initially been seen as components of effective care, they could also foster mistreatment and abuse. High profile scandals were allied to a recognition that overcrowded facilities often resulted in routinised and uncaring treatment that fostered a stigmatisation of mental ill-health (Goffman, 1961). The result of these diverse pressures was the decline and fall of the psychiatric asylum. This generated considerable challenges for governments and interest among academics. Geographers showed particular interest in the opening up of new spaces of care in the community for the treatment of the mentally ill, the process commonly referred to as deinstitutionalisation (e.g., Dear, 1978; Joseph and Boeckh, 1981; Smith and Hannam, 1981; Giggs, 1986; Moon, 1988; Kearns and Smith, 1993; Philo, 1997).

Closing Places and Opening Spaces

Thus far, we have traversed familiar and well-researched territory. There has been extensive scholarship on the history of the psychiatric asylum, asylum closure and the unfolding of the process of deinstitutionalisation. In contrast, our interest in this book lies in the under-explored territory of the post-closure fate of the asylum in the era of community care. We are interested both in the survival and re-framing of the idea of asylum and in the fate of asylum buildings and grounds. Our interest in this gap in knowledge emerged from our different but converging research concerns with deinstitutionalisation – work on the policy and human dimensions of asylum closure (e.g., Joseph and Kearns, 1996) and calls for the rehabilitation for the idea of asylum (e.g., Moon, 2000). These twin threads of interest were brought together by an opportunistic encounter with a surviving private-sector asylum in Canada (Joseph and Moon, 2002).

We see deinstitutionalisation as the basis for two challenges to the asylum: how to set up networks of community care and housing for former asylum residents; and what to do with large institutions, often with architecturally significant buildings on extensive, landscaped estates now deemed surplus. The former, patient-focussed challenge almost totally eclipsed the latter, institution-focussed challenge, both in terms of policy response and academic commentary. Notwithstanding the generally-accepted merits of moving to a community-based modality of care, contemporary accounts of deinstitutionalisation across various jurisdictions painted a picture that was anything but pretty (e.g., see Dear and Taylor 1982, Dear and Wolch, 1987). It proved easier to empty asylum wards and villas than to develop networks of community-based housing and care. The scale of the challenge comes into sharp focus when one thinks in terms of the 'reverse economies of scale' that were being encountered. Consider, for instance, the challenge of coping with the displacement of patients from even a relatively small psychiatric asylum, one with say 500 patients. A sub-set of patients, perhaps even a sizable one, would be deemed unsuitable for community care because of the severity of their illness. Places for these patients would need to be found in (expanded or newly created) psychiatric wards in general hospitals and provision made in budgets for their care and treatment. Assuming that one-fifth of patients fall into this category, there would remain 400 to be housed and cared for in the community. With say 10 residents as an ideal size for a group home, municipal planners would be tasked with finding 40 'suitable homes' in 'suitable neighbourhoods' (Smith and Hannam, 1981; Moon, 1988).

With respect to the residential dimension of care, the well-documented Toronto situation in the 1970s and early 1980s was probably typical (Dear and Taylor, 1982). Municipal officials and advocates of community care tasked with setting up group homes found neighbourhood residents to be less than welcoming, as reflected in the catch cries 'not on our street' and 'not in my back yard', with the stigma of mental illness following affected individuals from the asylum into the community (Dear and Taylor, 1982). Planning frameworks were found to be glaringly inadequate for the challenge posed by deinstitutionalisation (Joseph and Hall, 1985). The result in Toronto, as it was in many other cities in North America and elsewhere, was the concentration of residential facilities in deprived, largely inner-city neighbourhoods permissive of 'boarding house-type functions' but arguably bereft of the supportive therapeutic environment imagined as ideal by advocates of community care. The aftermath of these planning conflicts were imprinted deeply in the lives of former asylum residents relegated to 'landscapes of despair' or 'service-dependent ghettoes' (e.g., see Dear et al., 1980; Kearns et al., 1987; Kearns, 1990; Laws and Dear, 1988).

Our own collective engagement with this theme sought to re-incorporate the asylum into the critique of community care. Joseph and Kearns (1996) examined the then impending closure of Tokanui (Psychiatric) Hospital in the Waikato Region of New Zealand's North Island. They considered the social and cultural costs associated with the closure of a psychiatric asylum – the 'loss' of symbolic identity as well as employment and the abandonment of asylum-based innovations in the culturally-sensitive treatment of mental illness among Māori. In a subsequent paper (Joseph and Kearns, 1999), they provided evidence for the re-criminalisation of mental illness. A similar theme was pursued by Moon (2000). Drawing on a discourse analysis of former contemporary mental health care policy in the UK, he was able to demonstrate that popular understandings of mental health and well-publicised failures of community care contributed to a revalorisation of the asylum. In this framing, it was regarded as a place where people with mental health problems could be treated safely and separately, away from a public increasingly concerned with the risks and dangers posed by the delivery of care in the community.

Despite these small incursions pointing to the continued relevance of asylum facilities and the idea of asylum in the era of community care, there remains little work on the fate of asylums. Discourse on mental health care has referred often to notions of survival (Parr, 2008; Pinfold, 2000): patients *survive* their illnesses and patients and those charged with their care *survive* the (often chaotic) processes of deinstitutionalisation. But what of the institutions? What was their fate? Was there a 'life' for them beyond closure? And what happened to the notion of asylum that had so dominated thinking about the treatment of mental illness for more than a century?

Developing an Approach

The very sparse international literature on the uses to which former asylums have been put is of two general types, the first presenting a snapshot of re-use in particular jurisdictions (or parts thereof) at particular times (Dolan, 1987; Lowin et al., 1998; Chaplin and Peters, 2003) and the second presenting case examples of re-use (Franklin, 2002; Maachi, 2003; Joseph et al., 2009; Kearns et al., 2010; Bowden, 2012; Kearns et al., 2012; Joseph et al., 2013). This relative lack of prior attention to the subject matter of this book underlines the gap in knowledge that we seek to fill.

Notwithstanding the sparse prior literature, we found the three survey papers to be particularly useful in shaping our research strategy. In a US survey, Dolan (1987) sent questionnaires to 258 state hospitals, asking about changes to the size of grounds and buildings, 1970–1985. He found that 32 per cent of hospitals had undertaken property transfers involving 370 buildings and nearly 10,000 hectares of land. Of the new uses, 26 per cent were related to mental health care and 11 per cent involved correctional activities (e.g., prisons and juvenile detention facilities). Most of the other cases of re-use (53 per cent) involved activities such as recreation, education and housing. A decade later, and speaking to the results of a comprehensive survey conducted in 1996 of 206 large (> 100 bed) UK psychiatric and learning disability hospitals, Lowin et al. (1998, p. 129) reported that more than half of the sites made available through hospital closure were vacant and that "re-used land was most commonly deployed for agricultural, residential, education, leisure, business and other NHS activities". A few years later, the emergence of residential development as a favoured re-use of former psychiatric hospital in the UK was coming into focus. Chaplin and Peters (2003) surveyed 71 hospitals in six areas of England to determine the proportion of hospitals still open and the fate of those that had closed. Preserved buildings were found on more than a third of the sites, often as part of 'luxury' housing developments. Indeed, the authors reported that six developments were "entirely private with no public access, often guarded by security guards" (Chaplin and Peters, 2003, p. 227).

Based on these surveys of the 'fates' of former psychiatric asylum sites, we posit a four-fold heuristic framework for our study. The fates within this framework are not exclusive categories; they can overlap and recur. Despite the ostensible demise of the asylum modality, the first fate is *retention*. Mental health care uses such as outpatient clinics or small residential facilities may remain on site; health care administrative functions may also persist or indeed shift to a former asylum site in tacit recognition of its locational advantage, ownership and historical associations. In direct opposition to retention lies *dereliction*, a second fate wherein sites and buildings constitute an unrealised asset, abandoned upon closure, with the ravages of time, weather, vandalism, neglect and infestation by animals and plants becoming increasingly evident. Short-term dereliction may precede two further fates. Some former asylum sites may be *transinstitutionalised*: there may be a recognition of the amenity value of the grounds and of site accessibility such that the architectural shell and the site of the closed asylum (especially in urban locations) is converted into another institutional use such a prison or a tertiary education campus. With this fate we borrow and reformulate a concept generally used to refer to the cycling of clients through community care providers in both the health and criminal justice sectors (Prins, 2011). In other cases, sites and buildings may be converted to *residential uses* through housing developments in converted asylum buildings or through new build on what otherwise amounts to a 'brown field' site. In some of these cases, the bounded character of the asylum lends itself to the development of 'gated communities' where intruders are kept out where, formerly, patients were kept in. Both the transinstitutional and the residential fates may entail complete or partial demolition of asylum buildings as a precursor to redevelopment after a period of dereliction.

Turning to the fate of the idea of the asylum, our argument is two-fold. First, and in a direct way, the private sector in health care has continued to offer asylum to those who can afford its services. While in a sense complementary to the community care model pursued in the public sector, the persistence of a residentially-based modality in the private sector has served as a reminder of the enduring public support for notions of psychiatric asylum. Second, the original links between asylum, therapeutic landscapes and seclusion have a powerful contemporary resonance. They have been deployed both as a tool for the ongoing marketing of care in the private sector and as a selling point underpinning the re-cycling of former sites, notably for housing purposes.

We use a suite of case studies to anchor the development and illustration of our conceptual framework and key arguments (Yin, 1989). Our case study approach allows us to consider process without separation from context and provides a lens through which to identify key cause and effect relationships as well as to distinguish the exceptional from the normative. It also allows us to build on and enhance the limited existing literature cited above. We draw our examples primarily from three countries – Canada, New Zealand and the United Kingdom. For the most part, these countries have been on convergent paths in terms of the evolution of attitudes toward mental illness and treatment modalities. The evolution of mental health care in New Zealand was, for instance, heavily influenced by approaches in Scotland (Brunton, 2011). In Canada, developments combined the thinking from the United Kingdom with innovations, especially in the design of asylum buildings, from the United States (Paine, 1997).

Against this backdrop of considerable shared experience, we note significant dimensions of divergence that insert important aspects of difference into the case studies. One of these is the presence of (Māori-European) bi-culturalism in New Zealand, which pervades all aspects of health care delivery in that country (Durie, 1999). Another, and perhaps an even more defining attribute of the New Zealand context, is that deinstitutionalisation – the closing of places (of asylum) and the opening of spaces (of care in the community) – was considerably delayed (Brunton, 2003). In New Zealand the process occurred some 25 years later than in either the United Kingdom or Canada. The fact that this transition did not occur earlier, in the 1980s or even the 1970s, is attributable to the piecemeal approach of the New Zealand government to mental health care. Indeed, Hall and Joseph (1988) go as far as to label the government stance on mental health care in those decades as 'non-policy'; the lack of even a weak policy framework for deinstitutionalisation guaranteed an *ad hoc* approach to community care and the survival of asylums as the dominant modality of care into the 1990s. Delays in pursuing deinstitutionalisation meant that the policy debate on the closure of psychiatric hospitals and the opening of new spaces of care in the community was overtaken and engulfed in New Zealand by the ideologically-driven restructuring of the welfare state (Joseph and Kearns, 1996). There was no such conflation of policy directions in the United Kingdom, where successive waves of closure proceeded in the 1970s and beyond, notwithstanding local opposition focussed on the loss of employment opportunities and suspicion of the community care option (Jones, 1993).

A second important point of divergence relates to the dramatic shift in the role of the state in the provision of all aspects of health care in Canada in the 1960s. In that country, a growing and increasingly radical critique of asylum-based care (Dear and Taylor, 1982) coincided with a groundswell of support for the introduction of socialised medicine (Vayda and Deber, 1992). In contrast, in the UK and New Zealand there was little challenge to the persistence of well-established mixed public-private health care provisioning in the 1960s. Thus, in Canada thinking about the future role of asylum-based care in that decade and beyond was coloured by issues of hospital ownership and access to the new stream of government funding for all types of health care. Similar debates about the role of the public and private sectors characterised the United Kingdom following the election of a reform-minded Conservative government in 1979. In this case, the situation was one in which the dominant socialised model of health care was challenged by both the rehabilitation of the private sector generally and the introduction of fiscal measures enabling the growth and success of private care providers, most significantly in the care of the elderly but also in mental health care (Pilgrim and Rogers, 2001)

Other salient aspects of national context will be drawn out later in connection with specific analytical themes, but one final general attribute is worthy of immediate comment. New Zealand, with a population of a little over 4.5 million is by far the least populous of the countries from which we draw our case studies, with obvious implications for the scale and number of health care facilities. Consequently, by including New Zealand case studies in the examination of virtually every theme addressed, we 'over-sample' from that country. However, this brings with it the benefit of being able to generalise more definitively about the New Zealand situation than that in the other two countries, Canada and the United Kingdom, from which we draw the majority of our case studies.

In choosing case study institutions we had to draw boundaries without sacrificing the possibility of learning from parallel situations in which ideas that underpinned institutions and the physical fabric of those institutions were recycled. We have set aside from consideration more ephemeral treatment facilities. In all three of the countries introduced above there were, for example, residential institutions set up during each of the two world wars of the twentieth century to deal with service personnel with various physical and mental health problems. Such facilities were short-lived and small in scale compared to the major psychiatric asylums and did not have a comparable presence in the cultural landscape. Our focus has been on major established and long-lived facilities that were, and in some cases still are, fully integrated into the landscape of mental health care provision in the respective countries.

It is their visual presence and strong, almost visceral, cultural identity that makes these facilities such a distinctive example of the nineteenth century proclivity to establish networks of institutions to deal with health and social issues. The list of contemporaneous institution-building initiatives is both impressive and daunting – sanatoria, homes for the physically disabled and for the intellectually disabled, workhouses (although these pre-date the period under discussion here), industrial schools, facilities for the assimilation of particular ethnic groups, isolation hospitals and orphanages – to name but a few. However, it was only the asylum that combined three important sets of characteristics salient to our study. First, and as mentioned earlier, the asylums were large in number, built on an impressive scale and sited in large, often park-like estates. Second, they were long-lived and developed strong linkages with their local communities, through employment as well as service provision. Third, and arguably of greatest significance, the stigma attached to mental illness and to the sites of its treatment was (and still is) of a nature and intensity arguably matched only by prisons. This stigma seems resolutely to transcend time and place, in part because of the prominence of gothic horror images of the asylum in popular culture (see Chapter 7). We will acknowledge (and unpack) the various and several impacts of this stigma on the memorialisation, remembrance and re-use of individual institutions and on the re-deployment of the concept of asylum.

Notwithstanding our focus on the psychiatric asylum and on related developments in mental health care policy and delivery, we selectively draw insights from studies of the re-use of other types of institution, such as those noted above, and from scholarship which probes the interconnection and inter-penetration of the built and cultural landscape. By inter-penetration we refer to the processes by which aspects of the built landscape are given shape and meaning culturally, but which then, by their very existence (or persistence), transform the cultural landscape. This stance leads us to different literatures and constituent concepts, covering a diverse range of topics such as the re-use of contaminated brown-field sites, the adoption of 'asylum-related' imagery in the marketing of gated communities, and urban exploration. We hope that, in turn, our insights into the re-use of the former psychiatric asylum will inform theorisation in those fields of inquiry.

Structure of the Book

The remainder of the book is organised in seven chapters. Chapter 2 sets out the major ideas upon which our understanding of the recycling of the psychiatric asylum is built, noting first the overarching importance of policy. We then observe the difficulties associated with defining 'closure' for particular institutions (and associated challenges of developing closure narratives) and the complex ways in which the asylum and the layers of stigma associated with it can be deconstructed. We also point to the importance of landscape, both as metaphor and as the medium in which traces of the psychiatric asylum are preserved through heritage conservation. Specifically, we introduce critical ideas associated with memory, focussing initially on the concepts of memorialisation and remembrance. We then go on to introduce two further ideas that we have developed in association with our investigation of the recycling of the psychiatric asylum – *strategic forgetting* and *selective remembrance*. Following a discussion of the interplay of time and space in the recycling of the idea of asylum buildings and sites, the chapter concludes with a short discussion of the methods employed in our research, both generally in relation to our chosen strategy of case study research and in connection with particular lines of inquiry.

Chapters 3 and 4 deal with two aspects of survival. In Chapter 3 we consider the persistence of the idea of asylum in an era of community care and highlight the role of the private sector in its contemporary re-formulation. Case studies of the Homewood Health Centre (Canada), Ashburn Hall (New Zealand) and the Priory group (United Kingdom) are used to illustrate survival and business strategies. Particular attention is paid to the importance of marketing generally and to the specific use of 'imagineering' to re-work, revitalise and re-deploy notions like therapeutic landscape to sell the concept of asylum. We note that such strategies blur the boundaries between sites of mental health care delivery and facilities such as health spas and sanatoria, both in terms of the business strategies employed and the celebrity culture sometimes invoked. Chapter 4 considers the persistence of mental health care activities on former asylum sites retained within the public sector. We consider the retention of specific aspects of care and the balance between full and partial retention. Where core (residential as well as out-patient) psychiatric services continue to be delivered on former asylum sites, efforts to escape the long shadow of the past are noted in initiatives to re-name and to obliterate iconic buildings or other reminders of the past. Case studies consider the journey to retention in the UK, using the example of St James' Hospital, Portsmouth. This example is complemented by a consideration of the re-emergence of residential care as a treatment modality in Ontario and the persistence of 'forensic care' in all three of our case study countries. Forensic care caters to people at the interface between the health and criminal justice systems and is generally highly stigmatised. As a novel twist and coda to Chapter 4, we touch on the commemoration of the asylum era through on-site museums, with Porirua (New Zealand) and Glenside (United Kingdom) offered as examples. We connect this psychiatric museology back to the theme of survival by noting that the continuation of mental health services on former asylum sites itself constitutes a living memorial to the heritage of asylum care.

Chapters 5 and 6 present case studies of the recycling of asylum sites into other institutional uses and residential uses respectively. Tertiary-level education is used to illustrate the mechanisms involved in the asylum fate we term transinstitutionalisation. Two case studies, Lakeshore Hospital/Humber College

(Canada) and Carrington Hospital/Unitec (New Zealand) are considered in detail, with additional insights drawn from Challinor/University of Queensland (Australia) and Kalamazoo/Western Michigan University (USA). Strategies to 'purify' space and mask the former asylum use – which we refer to as strategic forgetting – are examined. Examples of deliberate memorialisation are set against the dominant trope of distancing from the past. In Chapter 6, in the examination of the recycling of asylum sites and (sometimes) buildings for the residential fate, we extend the consideration of strategic forgetting through illustration of complementary practises of selective remembrance associated with re-naming and re-imagining in marketing. Drawing primarily on three examples – Graylingwell and Knowle (both in the United Kingdom) and Sunnyside (New Zealand) – we examine adherence to heritage designations and land-use controls as a key tension. We describe how the peripheral (and attractively landscaped) sites of many asylums have led to their appeal for housing and transformation into (paradoxically) gated communities. We round out this consideration of what is emerging as a dominant re-use of asylum sites by considering parallels and contrasts with the re-use of military and industrial sites and infrastructure.

Chapter 7 deals with the significant number of former asylum sites that remain in a state of dereliction, our remaining fate. We consider the literal and metaphorical isolation of these abandoned sites and reflect on their presence in a postmodern landscape as spectral reminders that evoke memories and images of what was once mainstream and is now 'other'. We note the perverse attraction evoked by dereliction, enhanced sometimes by the conversion of former asylum buildings into 'horror attractions' but almost always underlain by a fascination with the imagining of the gothic asylum. Recent scholarship on haunting and spectral geographies is used to frame an interrogation of discourses on the derelict psychiatric asylum evident in the 'blogs' of individuals committed to the exploration of abandoned buildings and landscapes. Analysis of 'bloggings' for a number of former psychiatric asylums, including some of those used as exemplars in earlier chapters, reveals the identification of derelict sites as not only places of danger and discovery, but also as places where particular images of the psychiatric asylum are recovered and remembrances created.

Chapter 8 offers a series of conclusions concerning the fate of the idea of asylum and of its former sites. We see in the creeping distrust of the state-promoted panacea of community care an opportunity for at least the idea of asylum to be rehabilitated. However, we wonder whether such distrust is sufficient to transcend the stigma that still weighs down the memory of the psychiatric asylum. In terms of the re-use of former asylum sites and buildings, we reflect on the importance of location within the context of an urbanism that increasingly values brown-field sites and has exhibited a willingness to manipulate, if not accept, a challenging history of institutional use. We note that recent successes in combining heritage conservation with (profitable) housing re-development suggest that a number of former sites currently derelict may soon be swept up by eager developers and that intensification and/or diversification may occur on sites redeveloped in earlier decades. We also suggest that the overarching reality is one of mixed re-use in which paradoxically the re-appearance of residential mental health care will play a part. Where such mixed re-use does not occur, we believe that the on-going re-development of sites and buildings will make it ever more difficult to discern the traces of the asylum in the physical landscape, such that popular culture, with its interest in imaginings as much as memory, may be more important as a source of long-term remembrance of this previously-dominant mode of mental health care delivery than any physical traces in the landscape.

Chapter 2

Researching the Psychiatric Asylum

This chapter introduces and elaborates upon the ideas that shape our research; ideas which come together as an interpretive framework to guide the case studies that constitute the empirical backbone of our investigation. We draw on diverse literatures to develop these theoretical underpinnings, noting our own conceptual contributions. This mapping of ideas is set out in seven parts, and is complemented at the end of the chapter by a discussion of the research methods employed in our case studies.

We begin with the policy dimension. Our focus initially is on the overarching theme of deinstitutionalisation – the closing of asylum places of care and the opening of alternative spaces in the community – but we also reflect on the often overlapping policy shifts within and beyond the health sector that have influenced the outcomes of deinstitutionalisation. We argue specifically for the significance of overarching trends around the restructuring of the welfare state and the changing role of the private sector that have been identified generally as critical drivers of change in health service provision (see Blank and Burau, 2006).

Second, we examine the nature of asylum closure and its outcomes as the precursor and incubator of considerations of the re-use of asylum sites and buildings. We note, for instance, the difficulty of stating a date of closure for many psychiatric asylums, and suggest that closure should be seen more as a process than as an event, sometimes extending over several years or even decades. The resultant uncertainty concerning the future of institutions often carried with it important implications for the consideration of re-use. Our discussion of closure leads directly into a consideration of stigma. In deconstructing this third and critical concept, we highlight the various layers of stigma associated with mental illness, those affected by it and their places of treatment.

The remaining four sets of ideas relate to linked aspects of location, landscape and remembrance. We begin by examining location and landscape, treating the latter both as a metaphor and as the medium within which traces of the asylum are preserved. Particular attention is paid to the concept of therapeutic landscape – at first glance a concept that seems self-explanatory but one that, on reflection, is clearly amenable to idiosyncratic interpretation and even manipulation. These possibilities are hinted at by oppositional notions such as 'landscapes of despair' (Dear and Wolch, 1987), but they can be seen more directly in the therapeutic origin of the notion of asylum and its contemporary redeployment as a marketing device in the private sector. We then turn to an examination of the psychiatric asylum as heritage. The dynamic and complex relationship between the built environment and cultural norms and preferences is noted, and pointers are made to the importance of built form for understanding past cultural practices. Next, we introduce critical ideas associated with memory. The concepts of memorialisation and remembrance are considered, and two ideas developed in the course of our research are introduced: strategic forgetting and selective remembrance. To round out the development of our interpretive repertoire, we reflect on the interplay of the above ideas in time and space.

In Chapter 1 we stated our intention to use a case study approach to illustrate how the ideas introduced above play out in different places at particular times, exhibiting both commonalities and exceptions. Thus, the final section of this chapter reflects on our research methods. Our chosen approach was guided by several critical considerations. We needed to be sensitive, for example, to different national and local circumstances, including the availability of published information and concerning ethical issues relating to confidentiality, as well as varying protocols regarding access to sites and documentation. In this overview, we focus primarily on the two groups of methods that we deployed to varying extents in all of our case studies. First, we recount our use of visual ethnography, to assess site changes, building conditions and the presentation of the asylum, including the memorialisation of past use. More often than not, this amounted to a 'forensic' approach to fieldwork. Second, we discuss our use of documentary, pictorial and cartographic evidence and media reports to develop narratives of events and contrasting viewpoints, progressing from closure, through various considerations of re-use, to the present. More detailed discussion of sources and analytical strategies is integrated as necessary into each of the case study chapters.

Conceptual and Theoretical Underpinnings

Policy – Looking Beyond Deinstitutionalisation

Policy is the first building block for our interpretive framework and serves to guide our integration of other themes. In Chapter 1, we identified deinstitutionalisation as the key policy that gave rise to the closure of asylums and noted key drivers. Overcrowding, escalating maintenance costs, growth in reported cases of ill-treatment, persistent stigmatisation, increasing recognition of the dangers of institutionalisation and the emergence new therapeutic developments have all been noted as contributing factors in the move away from asylum care. Deinstitutionalisation as a policy is not, however, the only imperative that can be implicated in the retreat from the psychiatric asylum as the dominant modality for the treatment of mental illness. Rather, an ideologically-driven process of restructuring was, to varying degrees, also evident as an impetus to institutional closure in our case study countries.

In New Zealand, for instance, it is evident that the large state-funded Tokanui psychiatric facility in the central North Island closed under circumstances in which the principles of deinstitutionalisation became subsumed within a neoliberal logic of restructuring (manifest as public sector fiscal restrictions, a culture of 'value for money', and calls for greater accountability) (Joseph and Kearns, 1996). The net result was that treatment modalities were reduced to their organisational and financial logics while humanistic considerations, such as the ruptured sense of home for long-time residents, were overlooked. This shift finds parallel in the move to patients being viewed as a source of risk (Warner, 2006), with mental health professions becoming more interested in sets of 'risk factors' rather than in the ill-health of individuals or groups, with the consequence that "there is no longer a subject" (Castel, 1991, p. 288, emphasis in original).

The beginning point for risk assessment, according to Castel (1991), is identification of the dangers to be prevented rather than direct experience of some kind of threat based on contact with an individual. In the new era of community care, patients assessed as 'high risk' were henceforth detained in new high security facilities, often located adjacent to hospitals (McCallum, 2008). This weighing up of the factors deemed liable to produce risk involves what Castel (1991, p. 282) describes as – "a transition from the clinic of the subject to an 'epidemiological' clinic" in which specialist medical carers invariably defer to administrators who, in post-asylum times, often assume autonomy in allocating individuals into categories of risk (Moon, 2000). Priority in planning post-asylum services was invariably placed on the retention of existing secure units or the building of new ones. For example, work by two of the authors on deinstitutionalisation in New Zealand in the 1990s highlighted repeated delays in establishing community-based residential and treatment facilities in the Waikato region, in contrast to the early planning and development of a regional forensic psychiatry unit (Joseph and Kearns, 1999). The containment of risk trumped the arguably greater challenge of housing the majority of former asylum residents not deemed to be dangerous.

In our other case study countries the encounter between deinstitutionalisation and restructuring had some similarities but also differences. In the UK the initial phase of deinstitutionalisation took place in the 1960s and early 1970s in the twilight years of the Keynesian welfare state. A strong state with a commitment to public expenditure ensured that the early phases of the closure process proceeded at a measured pace. As the UK economic crisis of the mid 1970s hit, pressures changed. Fiscal restraint and severe cost pressures within the NHS emerged as an imperative at the same time as the implementation of the process of deinstitutionalisation reached a stage when reduction of inpatient numbers gave way to asylums actually closing. Much of the subsequent history of closure has reflected the continued difficulty of reconciling the ideals of deinstitutionalisation with the prevailing political-economic imperative for cost containment. At the national level in Canada the early phases of deinsitutionalisation proceeded at the same time as the provinces were assuming the ownership of general hospitals and sole responsibility for the delivery of health care. This commitment to state funding of health care, enshrined in the Canada Health Act (1967), coupled with stronger government control via ownership of general hospitals, created the opportunity to begin the process of closing provincial psychiatric hospitals. Thus residents of psychiatric asylums could be transferred in the 1970s and 1980s to newly expanded wards in general hospitals in advance of the development of a fully-fledged community care system. The impact of public sector retrenchment and cost containment came later, in the 1990s, by which time asylum facilities had largely closed but new spaces of care in the community were still poorly developed (Sealy and Whitehead, 2004).

In most jurisdictions, the net result of policy shifts towards community care was a network that, again drawing on Castel (1988), arguably comprised a more subtle and spatially dispersed means of overseeing and regulating people with mental health problems than the asylum. Through its diverse forms of expert intervention, community care facilitates monitoring people with mental health problems in ways which are arguably as comprehensive and committed to surveillance as the asylum system before it. However, this surveillance has moved from something done solely by professionals 'behind walls' to something done by ordinary citizens and the media as well as physicians and social workers.

Restructuring in mental health care has thus been underpinned by what we might broadly label the 'neoliberal turn'. This shift away from the early ideals of deinstitutionalisation has increasingly shaped the context for both the closure of large public institutions and the survival of private residential mental health care. Its hallmarks have been the encouragement of choice, contracts and competition as well as reductions in public expenditure. Across a range of health and social care sectors that go beyond mental health to include elder care and disability, a public/private mix in the funding, delivery and regulation of care has become increasingly acceptable (Bartlett and Phillips, 1996; Bigby, 2002). Additionally, the promotion of client choice ensures that people who desire and can afford private hospital and residential care will have appropriate opportunities available to them (Kearns et al., 2003).

With respect to the closure of psychiatric asylums, this neo-liberal context has had three consequences. First, it has hastened the closure process to save money tied up in the maintenance and repair of asylum buildings, trusting in competition to keep costs down in a contract-based community care system. Second, it has complicated the map of successor community care provision by a mix of public, private and voluntary actors, making navigation of the system difficult for clients who had been used to a more unified institutionalised care system. Third, asylum land and buildings have been commodified as assets with value in the market that can be realised through sale. The realisation of such value was attractive to governments searching for ways to finance new facilities and services or, all too often, to reduce accumulated levels of debt.

In the case of the survival of private sector residential mental health, it is the choice aspect of the neo-liberal turn that has assumed importance. The adaptability of private residential mental health care provision and its ability to serve niche markets is underwritten by the idea that people should be free to choose private-sector provision. The private sector can also offer flexibility to the public sector by offering a modality that is no longer deemed to be a priority by government, but is nevertheless desired by a sub-set of public-sector clients (Knapp et al., 1997). Further ideas, such as interdependence and mutuality, by which the public sector contracts out to private providers of residential care, are examined in the next chapter.

Shifting from a national theoretical and policy context, to a more local perspective, it follows that in any single jurisdiction the cessation of asylum functions needs to be seen as a local implementation of a wider spectrum of changes. Such changes are invariably associated with economic and service-sector restructuring as well as, in some areas, significant local demographic shifts (Kearns and Joseph, 1997). In areas of limited or no demographic growth, hospital closures loosely coincided with a range of elements of the service infrastructure being 'downsized' or completely removed, resulting in a surfeit of opportunities for investors willing to re-purpose buildings and sites. Such was the case in New Zealand, where amenities previously regarded as core to communities (e.g., the local bank, chemist, post office) became a memory in many small rural towns (Joseph and Chalmers, 1998). These closures, occurring from the mid-1980s onwards, made sites and buildings available for re-use that, compared to asylums, were generally smaller in size and more centrally located. Moreover, and as we will elaborate, they were bereft of the stigma bound up in the bricks and mortar of the psychiatric asylum. In contrast, in areas where population growth has been higher and economies have been more buoyant, the neo-liberal imperative has seen closed asylum sites gradually recast as opportunities for expansion, development and investment. This recasting has been particularly rapid in urban settings in the UK.

Closure Narratives

The second theme in our construction of an interpretive framework involves closure narratives, which we contend act as the precursor and incubator of considerations for re-use. While the term 'closure' speaks most often to a temporally-specific moment of change, in this section we consider the difficulty in stating categorically the date of closure for many psychiatric asylums. We suggest that like the enveloping phenomena of deinstitutionalisation and restructuring, closure should be seen more as a process than a historical moment. Indeed, the protracted nature of the 'real' as opposed to 'formal' processes of closing psychiatric asylums served in most instances to dilute and deflect public debate (as well as media interest) in the demise of these previously-prominent institutions.

As with factory closures, the process of terminating asylum functions often involved a 'run down' of capacity that began well in advance of the actual announcement of closure, such that in their final years asylums were reduced to shadows of their former selves, both in terms of the range of services provided and number of patients in residence (Joseph et al., 2009). Frequently, the physical condition of buildings was allowed to decline during a 'wind-down' period, starting well ahead of discussions of closure. An emergent state of decrepitude might lead to further decline: why invest in a facility that is regarded as not only outmoded, but also in poor condition? Concern with the increasing cost of maintaining ageing infrastructure generally pre-dated calls for deinstitutionalisation, but clearly the latter cast a shadow over decisions on both upkeep and closure. Not surprisingly, the state of buildings also affected the future use of hospital sites. By way of example, the Danvers State Hospital in Massachusetts experienced a particularly prolonged closure process, with budget cuts constraining maintenance in the 1960s and leading to a retreat from some wards. Through this incremental process the majority of the hospital was abandoned and falling into disrepair by the mid-1980s. However, it was not until 2004 that the entire hospital was formally declared closed. At a smaller scale, the poor condition of the former Seaview Hospital in the South Island town of Hokitika, New Zealand, was cited as one of the reasons for opposition to its purchase by members of a local Māori group, while litigation over the sale of Kingseat Hospital south of Auckland focussed on the failure of its previous owners to provide an accurate account of its condition to prospective buyers (Joseph et al., 2009).

Closures were variable, then, in their degree of finality. In many instances, there was an incremental decanting of patients, with selected mental health care functions remaining, temporarily or indefinitely, on some sites. These persisting service functions were often re-cast as 'clinical centres' or 'mental health villages' to bring them in line with the new era of community care. This retention of mental health care services on the site of former asylums is considered in detail in Chapter 4. In addition to the state of disrepair mentioned above, relations with the local community and the attitudes of government agencies also affected the speed and style of closure processes. Local communities often expressed worries about the loss of services and employment, while concerns regarding the need to implement community care and contain costs were a key focus for government agencies (Joseph and Kearns, 1999). Indeed, in many instances government agencies and local communities seemed almost to be referring to different institutions when expressing their particular concerns in often heated debates about the closure of particular institutions. What for some might be a welcome removal of a stigmatised service or a 'job well done' on the route to implementing community care, might, for others, amount to a loss of community purpose and identity (Kearns et al., 2012).

Stigma

The idea that asylums represent stigmatised services is well-established and central in both academic discourse and the popular imagination. It is the third component of our interpretive framework. For academics, following Goffman's (1961, 1963) seminal discussion of the concept, stigma is most commonly ascribed to people. Significantly, it is seen not only as a social construction applied to others but also a burden that those who are 'othered' assume in response to the 'affront' that is reflected back to them by way of peoples' attitudes or reactions. This approach has been criticised for being too vague and individualised in its attribution. In response, Link and Phelan (2001, p. 363) have redefined stigma as "the co-occurrence of its components – labelling, stereotyping, separation, status loss, and discrimination". They add that, for stigmatisation to occur, power must be exercised. It follows that stigmatised people may exercise their own power by asserting control over the very label that otherwise might be a burden. Indeed, the rise of contemporary cultures of pride and victimhood have sanctioned people accepting labels that might earlier have been regarded as stigmatising. What was once a 'badge of shame', therefore, becomes up-ended into a source of social entitlement and belonging: from stigma to 'mad pride'. Fitzpatrick (2008), for instance, points to 'bipolar disorder' as more acceptable than mere 'depression' and accruing sufferers both sympathy and attention. He also claims that people embrace being a victim of 'work-stress' for its implied marker of dedication and loyalty to a cause. Notwithstanding these contemporary trends however, for much of the period covering our study, mental illness has been feared and those suffering from it have been marginalised.

While stigma applies most readily to people, we argue that it applies equally to our interest in the places where people with mental illnesses were treated. Asylums became stigmatised locations. Even after residents of the former asylums had moved on – literally and metaphorically – the immobile presence of the sites of their former treatment remained discernibly present as a reminder of a stigmatised past. In the absence of people, much of the stigma loaded onto the sites of past treatment was amplified and also valorised by hyperbolic images drawn from popular culture such as the unforgiving institution with its uncaring staff portrayed in the film *One Flew Over the Cuckoo's Nest*. Just as the experience of stigma involves living with characteristics that are 'deeply discrediting' (Fitzpatrick, 2008), we also suggest a form of transference in which asylum sites and buildings take on a stigmatised and discredited reputation through former occupation and association. The contemporary re-purposing of asylums as elite housing estates, complete with the retention of distinctive architectural features, suggests a transmutation of the stigma associated with the built form of the asylum, via the recognition of heritage values, as a point of distinction. We consider this rehabilitation of a previously stigma-laden built form in detail in Chapter 6.

Looking across both the pre- and post-closure phases, we argue that there are two layers of place-related stigma at work. First, asylum sites accrue stigma through the mode of care that was the rationale for their buildings becoming tainted and out-dated (modality-based stigma). Second, and to a lesser extent, stigma develops as, one by one, the facilities themselves become de-valued, being regarded as redundant and insufficiently distinctive to constitute 'heritage' (facility-centred stigma). In both manifestations of site-directed stigma, some manifestation of the exercise of power identified by Link and Phelan (2001) is at work.

The stigma associated with the psychiatric asylum – the almost universal characterisation of it as an inhumane and outdated treatment modality – was accentuated in debates that occurred during the often protracted period between the announcement and the completion of closure. Such characterisation has generally remained salient in subsequent debates over re-use. Thus it is not surprising that as a built form the asylum can be seen to have an "ambivalent quality" … "symbolic of fear and oppression, but architecturally impressive monuments" (Franklin, 2002, p. 174). How does this stigma actually play out in the cultural landscape? It is helpful to remind ourselves that the roots of 'stigma' involve both palpable markers (stigmata) and tainted reputation. In the case of closed psychiatric asylums, we contend that it is tainted reputation that is the primary trace upon which more physical triggers of memory are dependent and which largely determine their interpretation.

The very names of (former) asylums signify this stigma and ensure its perpetuation. Following closure, changes in ownership frequently act as a key to new possibilities for a site, but do not guarantee an escape from the shadow of stigma. The fate of asylum sites can therefore be seen as reflecting Pred's (1984) notion of places as always in a state of becoming – a view alternately expressed by Anderson (2009, p. 52): "places are not nouns … rather they are verbs, they are doing things and always active". In the case of our focus, a closed psychiatric asylum is not a static edifice but rather speaks of a stigmatised past and resists a future until the traces of that stigma are addressed. One way of addressing this aspect of the shadow of stigma is by renaming the location. Historically, the names given to psychiatric asylums often conveyed positive, naturalistic environmental qualities or historical connection. As such, they reflected the norms associated with treatment through removal to sites of tranquillity, and they were, arguably, part of a broader effort to underline links between therapy and landscape (Gesler, 1992a; Moon et al., 2006). Over time these names accrued a darker meaning signifying stigmatised places. In the quest to create a distance from this past, re-naming has been undertaken. This process deploys names to create new norms (Berg and Kearns, 1996) and the 'stroke of a pen' serves, at least in official discourse, to sever links with the past (Kearns et al., 2012). Distance is established from names that are invariably seen to be a medium for the propagation of negative memory and emotion. Changing names offers the opportunity to apply new norms and replace what ironically were, in some cases, already euphemistic original names with new and equally banal names expressing alternative visions for the sites.

Location and Landscape

The fourth conceptual theme embedded in our interpretive framework is location and landscape. Philo (2004) writes of asylums as predominantly 'removed locations' in rustic surroundings away from densely populated centres, quite literally 'far from the madding crowd'. Such locations were conventionally regarded as having beneficial qualities flowing from the removal of stressors, immersion in 'natural' settings and the opportunity to engage in agricultural labour. At the time of their origin the 'setting apart' of asylums invariably meant their location was in rural or suburban landscapes that might be coded as 'idyllic'. Qualities such as fresh air and green spaces can be seen as amounting to informal 'therapy' generated by the physical context of the asylum setting and consciously augmented by the contrived community fostered by the social organisation of such places. In sum, these qualities comprised an attempt to create a sanctuary that was consciously set apart from what was regarded as the excessive stimulation of everyday urban life.

Of course, removed locations also had other, non-therapeutic attributes and even dangers, including the potentially negative impact of the spatial separation of people from family and friends. Spatial separation might encourage abandonment and isolation. These negative consequences of separation were later seen to lead additionally to further problems. As overcrowding became the norm, residents of spatially removed asylums succumbed more easily to institutionalisation, amounting at times to a debilitating state of 'learned helplessness' (Seligman, 1972). Moreover, spatial removal ensured that the asylum could become to some extent a hidden world; in this hidden world external awareness was reduced. For those inside the asylum there was reduced contact with the outside. Everyday life became a routinised norm and life outside a distant and disorganised memory. Therapeutic possibilities remained but could be subverted by the norms of the 'total institution' (Goffman, 1963). In the worst cases, seclusion might serve to hide abuse. Those outside the asylum often had little knowledge of what went on inside. Images might be idealised, stressing order and recovery through seclusion; equally they might dwell on the unknown and emphasise fears about mental illness. Through this latter route, separation came to foster stigma.

The creation of asylum spaces of care implied the imposition of boundaries. For their founders, this often took the form of walls of separation between the asylum and the surrounding community. The sharpness of such demarcations was somewhat softened by the park-like characteristics of the estate surrounding asylum buildings. In Chapter 4, we introduce the idea of 'boundary flexing' by which the zone of transition between surviving asylums and community has been encroached upon and new points of contact established. Additional examples of boundary flexing, involving real or symbolic separations, are offered in Chapters 5 and 6. In Chapter 7 we examine the case of derelict asylums and the phenomenon of urban exploration. In the latter practice, boundaries are seen as a challenge to be overcome as an integral part of the experience. In contrast to this literal interpretation of boundaries, in Chapter 3 we introduce a very different use of the concept. Specifically, we examine boundary crossing by which institutions move from what appears initially to be one polarised option to another, from the private sector to the public sector or from the institutional to the community modality of care. We term these shifts 'mission movement'.

Turning to landscape, we are particularly interested in the idea of therapeutic landscape, a concept that Gesler (1992a) introduced into health geography. To Gesler therapeutic landscapes include places or settings that have a reputation for healing founded in a combination of factors including historical precedent, natural attributes and symbolic association. The term has subsequently come to be recognised as a significant theoretical and now arguably overused construct within health geography (Kearns and Moon, 2002) that has captured the imagination of geographers as a tool with which to interpret places that have a reputation for healing or are health-promoting (see Williams, 1999; Wilson, 2003, Smyth, 2005). For us there is a risk of over-attributing therapeutic characteristics to the often rustic and bucolic settings of the asylum. As Conradson's (2005) critique reminds us, whereas the therapeutic landscape perspective has helpfully highlighted the environmental, social and symbolic dimensions of sites of health care and health-seeking behaviour, there has been less consideration given to the relational dynamics through which these therapeutic effects emerge. More significant challenges to the therapeutic landscape idea can, however, be seen in various alternative deployments of the term landscape which reflect ways that place can corrode wellbeing. These metaphorical uses range from the breeder hypothesis in which deprived inner-city settings are seen to generate mental illness among resident populations (Giggs, 1973), to the notion of 'landscapes of despair' (Dear and Wolch, 1987) where marginalised people, including people with mental health problems, are consigned and arguably confined to impoverished neighbourhoods. This distinction between therapeutic and non-therapeutic landscapes leads us to reflect how sites like asylums can simultaneously be therapeutic and detrimental to wellbeing.

Notwithstanding the contemporary ambiguities noted above, the original vision of the asylum in terms of location and landscape design appears, at least in part, to be predicated on ideas related to therapeutic landscape. Park-like grounds, seclusion and healing through removal from society and exposure to the positive properties of particular places were deeply embodied in traditional notions of the asylum as a care delivery modality. Despite this seeming resonance, historic asylums have not, to our knowledge, been recognised explicitly as therapeutic landscapes in the formal sense. There has however been tacit acknowledgement. Parr and Philo (1996) hint at the link between the historic asylum and the concept of places as 'therapeutic' and Philo (2004) has recounted how the historical justification for asylum care emphasised the virtues of offering a sanctuary that was intentionally removed from the (over)stimulation of everyday life in the rapidly changing emergent industrial city (Sennett, 1992; Simmel, 1995).

Though the idea of the therapeutic landscape may have been important in the genesis of the asylum, it was its negative counterpart, the 'landscape of despair' that came to be the predominant construct as the asylum era came to an end. Indeed, as Gleeson and Kearns (2001) contend, the implicit historic binary construct of 'asylum: good, community: bad' has, since deinstitutionalisation, been recast as 'asylum: bad, community: good'. This said, as the promise of community care foundered on the practicalities of funding, social support and public acceptance, this coding has seemed at times to be asylum: bad, community: bad. Indeed, to some, the asylum itself came to foster mental ill-health through a form of 'breeder' process in which confinement rather than treatment became a dominant characteristic (Busfield, 2011). In this way, and further enabled by ignorance and a general lack of public awareness and interest, the once-protective asylum transitioned in at least some cases into a landscape of isolation and despair. The legacy of this transition presents a major challenge to the marketing of present-day asylum provision in the private sector (see Chapter 3), to continued use of asylum sites for current health care purposes (Chapter 4) and to the re-purposing of former sites (Chapters 5 and 6).

Heritage

We now turn to a dimension of our interpretive framework that is not explicitly related to mental ill health and its treatment: heritage. One of our central concerns in this book is the question of what has happened to the land and buildings left behind after asylum closure. There are different dimensions to this question. In dynamic and complex ways, the built environment can both reflect and contribute to cultural heritage. Buildings can 'live on' as a key to understanding past cultural practises. Anderson's (2009) idea of 'traces' can help us discern connections with issues of heritage conservation. For Anderson, traces are both material remnants (e.g., buildings) as well as non-material consequences (e.g., events and emotions) with 'heritage value' being determined, in the first instance, by the way in which their original use revives recollections of past feelings for place. Such recollections can be, in Edensor's (2005a, p. 829) terms, affective in the sense that they evoke personal memories in contrast to the sanitised " ... fixed, classified, and commodified memories purveyed in [formal, designated] heritage and commemorative spaces".

As noted in Chapter 1, across the western world, concerns about the re-use of asylum infrastructure made newly redundant by governments embracing deinstitutionalisation came a poor second to a preoccupation with setting up community-based networks of care for people with mental health problems. The substantial infrastructure that became surplus to newly expressed needs stood almost forgotten as health planners and elected representatives turned their attention to the challenges associated with gaining acceptance of a community modality of care (Dear and Taylor, 1982; Philo, 2000; Wolch and Philo, 2000). Given the distinctive architectural characteristics of asylums, one might expect heritage conservation to be one outcome of closure. Consequently, we need to keep in mind that different types of re-use carry with them specific implications for the realisation of heritage conservation. For instance, it has been noted that redevelopment of asylum sites for housing is the re-use with most financial return and, especially in the United Kingdom, such redevelopment has been seen as a way of funding heritage conservation (Franklin, 2002). However, even when asylum buildings have been 'saved' for re-use as housing, there is often limited engagement with the past use: heritage designations invariably appear to facilitate the preservation of grounds and building facades but little more (Weiner, 2004). We probe this connection further in Chapter 6.

We speculate that the stigma associated with mental illness and its sites of treatment has worked to reduce the likelihood of proactive heritage conservation, even in instances where buildings are seen as emblematic of particular eras and design traditions. While some attention is given to identifying the works of major asylum architects and the preservation of park-like asylum grounds can attract public support, the sheer scale of asylums estates can often militate against the heritage listing of more than a few key exemplars. Though it might be possible to draw parallels with other 'brownfield' sites where a former use creates challenges regarding the preservation of 'heritage' (e.g., see Davidson and Lees, 2005; Edensor, 2005b; Hamnett and Whitelegg, 2007; Jonsen-Verbeke, 1999; Watson and Wells, 2005), there are also crucial distinctions in the asylum case. Brownfield sites pose many challenges, with land contamination being a major feature Stigma is an issue of a somewhat different order. On other types of brownfield site it is seldom encountered to the extent found in the case of former asylums. This situation raises fundamental questions about how memories are preserved or triggered by former land uses.

Memory and Memorialisation

Our interest in the material aftermath of deinstitutionalisation leads thus to the sixth theme in our interpretive framework: memory and memorialisation. Our interest in traces of past use on asylum sites has been informed by the work of Nora and his concern for how memory is embodied in sites "where a sense of historical continuity persists" (Nora, 1989, p. 7). We see 'historical continuity' to be the ongoing use of sites designated, or buildings constructed, in another time for another purpose. To Nora (1989, p. 9), while history is "always prosaic" and "universal", memory is "blind to all but the group it binds". In other words, mainstream historical accounts often aspire to be descriptive and dispassionate whereas memory is nuanced and affect-laden, connecting people whose lives have been shaped by common place-specific experiences. Neither are memories fixed and immutable. Landzelius (2003) argues for 'commemorative disremembering'. In the case of the former psychiatric asylums, this fluid process might mean that the understandings of present-day users about the meaning of an inherited site differ substantially both from the memories of former psychiatric inmates and also perhaps from the understandings held by those involved in the immediate purchase and conversion of the site from one use to another.

In a general sense the built environment of former asylums abounds with elements that evoke memories, whether personal or collective. The relict land use speaks directly of the past, particularly when contemporary budgets constrain the extent of renovation. A substantial literature testifies to the particular importance of collective memory in a range of settings. Sites of collective memory are likely to be kept 'on the map' and may be complete buildings, ruins on a 'heritage circuit' travelled by tourists, or strategically preserved, even purpose built – and often politically contested – memorials (Marshall, 2004). The links between landscape and memory are pervasive (Schama, 1995). Symbols and signage lend authority to particular readings of places, with memory being consequently "reassigned and controlled" (Gough, 2004, p. 251). This may be particularly the case in sites where there is a powerful collective memory such as places of battle and calamity, war memorials or concentration camps (Buntman, 2008; Charlesworth, 1994; Stenning et al., 2008).

Asylums appear to be somewhat different. While without doubt they can evoke a collective memory, it is generally a stigmatised one and, more importantly, one that tends to be individualised and personalised. Generally there has been a limited collective will to materialise remembrance in an overt manner. The memories (and traumas) are personal, indeed, often also associated with shame. We can further speculate that there will be an absence of memorialisation when new uses are in a formative state and the 'becomingness' (Pred, 1984) of the site as a place of alternative activity emphasises futures and new beginnings rather than a potentially stigmatised past. The process of memorialisation is therefore fluid and complicated by the passage of time (Burk, 2003).

We speculate that, beyond public memorials to past occupants, the traces of the psychiatric asylum that remain constitute not only material indications of a past modality of care but also an expression of the long shadow of collective and personal remembrance. This remembrance is triggered to varying degrees by the symbolic as well as material 'remains' to be found at now-closed sites and the positionality of those seeking to interpret the sites. A related question is whether the maturing of a new use or uses on former asylum sites will, in time, result in a coming-to-terms with the stigmatised past and greater memorialisation, or a complete forgetting. Those who acquire and re-develop sites are faced with a dilemma: to attempt to silence the past or offer 'too much memory' and risk unsettling their own aspirations.

Drawing on these ideas, we find it helpful to distinguish between the physical memorialisation of former uses and their remembrance. Memorialisation may have evolved over a considerable period of time. It may focus on notable events or people and thereby only indirectly celebrate the use itself. Remembrance, in contrast, can be triggered by memorialisation or may simply comprise narratives of past use (e.g., on the internet and in print media). Moreover, for us, remembering is re-membering: a putting-back-together of the past. In our case-examples of re-use and dereliction (see Chapters 5–7), this re-membering is inevitably partial, not only by virtue of contested memories clouded by differing notions of stigma, but also because only the buildings (or some of them) remain. The people, actions and performance of psychiatric care have gone. What once made the asylum and gave it life has been removed and we are left with sites haunted by absence. Into these spaces have sometimes flowed new uses with recast futures disturbed at the margins by echoes of a receding, partially remembered, and potentially disturbing past.

Building on our consideration of the foregoing themes, we have developed two related concepts to assist our work of interpretive analysis: strategic forgetting and selective remembrance. In terms of memorialisation and remembrance, Chaplin and Peters (2003, p. 228) report that while developers of former asylum sites often deployed nouns in their advertising – such as 'seclusion' and 'sanctuary' – that could be applied to the predecessor asylum uses, they very rarely made reference to those former psychiatric uses, "possibly reflecting the stigma of their former existence". Stripped of their original name and identity, and set, as they are, within extensive parklands, the casual observer could be forgiven for mistaking a recycled asylum for the buildings and grounds of a refurbished stately home. The past use has been strategically forgotten.

While strategic forgetting is very evident in accounts of the re-development of former psychiatric asylum sites for housing such as those provided by Franklin (2002), it seems that there is sometimes a companion strategy that involves 'selective remembrance' focussed on intrinsic components of the former use such as architecturally-distinguished buildings. To us, such selective remembrance serves as a counterpoint to the silences of strategic forgetting and represents an important dimension of the agency exercised by stakeholders in the re-development process. These complementary concepts interrogate the creative tension between the retention of positive memories of the asylum and the simultaneous obscuring of more negative aspects of that past.

Time, Space and the Asylum

We contend that the persistence, re-use or dereliction of a particular asylum cannot be explained without reference to the specific characteristics of the health care system to which it belongs and the interplay of its location and the timing of closure. We touched on the former in Chapter 1 and earlier in the present chapter; here, for the final theme in our interpretive repertoire, we consider the interplay of time and space in both asylum closure processes and the survival of the asylum idea.

Focussing on the issue of time, we have already noted that asylum closure processes could be very protracted. The significance of delays in acting upon announcements of closure could be two-fold, both during the run-up to closure and during the period between closure and a decision regarding a new use. First, possible opportunities for re-use could emerge and disappear. Second, and arguably of greater importance, buildings could deteriorate, sometimes to a considerable extent, increasing the cost of restoration and re-purposing. Such concerns make space for the politicisation of both the closure process and consideration of specific types of re-use. In particular, delays enable communities to mobilise in favour of specific issues such as heritage value, service retention and public safety.

Possible re-uses of sites and buildings are also circumscribed by broader changes in societal values and preferences, and these evolve over time. We offer two examples. First, attitudes towards what constitutes heritage change over time and this can have a bearing upon the propensity for representative buildings to be deemed worthy of preservation. We speculate that in the early stage of any wave of infrastructural closures, remaining examples will be sufficiently ubiquitous to reduce the possibility of retention. However, when examples become fewer and especially when they are deemed to be of architectural significance, there is greater likelihood that they will be culturally constructed as worthy of a heritage designation. At this time, sentinel examples will be preserved, sometimes at considerable cost and effort. A second example concerns shifts in social attitudes towards particular possibilities for re-use. Some replacement developments may not be immediately acceptable on closure. However, with the passage of time and growing public awareness of the full range of alternative re-uses or the perils of continued dereliction, a re-use originally spurned may become acceptable or even attractive.

Turning to location, as for any redundant infrastructure there will be spatial variations in opportunities for re-use. In general it is likely that relative proximity to large centres of urban population will be a key differentiator of potential re-use. Only for remote rural asylum sites in areas of population decline will there be no land market advantages to asylum closure. Urban asylum sites with extensive green space will, *a priori*, be intrinsically most attractive. They represent sites with pre-existing built development where further development permission may be more easily available. They also facilitate urban expansion though their location on the edge of cities. Finally, their grounds provide passive and active recreational opportunities.

Proximity to a former asylum is a key to understanding the interplay of these temporal and spatial themes. Such proximity has been associated with awareness of the stigmatised facility. Yet attitudes to asylums were not necessarily negative. For those living near to former hospitals, proximity may have been associated with positive perceptions that potentially contribute to place attachment. In these proximate communities, strong memories might be triggered and place attachment sustained, often at a very personal level, by even very subtle traces of the former use. We can speculate that such re-envisioning of the past, where there are now only suggestive remnants, will gradually erode over time with the passing of those who worked at, or resided in, these institutions. Proximity will thus weaken over time as a determinant of memory. A slightly different logic underpins the survival of the asylum in the private sector. Survival is facilitated when policy conditions and the legislative environment are permissive and proximity to markets provides a pool of potential clients.

Methods

The chapters that follow draw on a blend of field observation and interpretation of textual and visual materials. We explore the specific application of these research methods in each chapter and confine ourselves here to outlining the rationale and general approaches that we have employed in operationalising our chosen methods.

Our field visits took place over the period 2003–2014. The length of this period allowed us to make serial visits to a number of key sites in order to assess the progress of change and chart redevelopment as it took place. In other cases we made a single visit with the aim of checking complementarity and commonalities with the developments on our key sites. We also visited derelict sites that were only amenable to limited observation due to access restrictions. All visits entailed an assessment of the contemporary condition of sites and buildings and an investigation of the nature and extent of current uses and progress on planned redevelopments. Additionally we visited representative examples of contemporary private psychiatric hospitals in our chosen countries with similar research objectives.

In the course of each visit we developed a visual and written record of our observations, following what Rose (2001, p. 3) calls a critical visual methodology, in which we "think about the visual in terms of the cultural significance, social practices and power relations in which it is embedded". We took field notes and photographs of visual signifiers of former, current and projected uses. Some of these signifiers were obvious and easily identified; others required the exercise of more forensic fieldwork skills, seeking out hidden monuments and memorials. In most cases the process was straightforward; we found, or were guided to, particular locations and openly recorded our findings. A few cases were more challenging. In environments where there were continuing health care uses, we had to be sensitive to privacy requirements and avoid photographing clients; we were also more closely chaperoned on such sites. On derelict sites other challenges prevailed. Alongside issues of access we had to be attentive to health and safety requirements. While we would hesitate to describe ourselves as 'urban explorers' (Garrett, 2014), there were times when we accessed derelict environments similar to those sought out by such individuals, experiencing some acutely memorable transgressive moments as we stepped over locked gates and into derelict spaces.

Both as a precursor and as a follow-up to our site visits, we accessed contemporary written information on the use and re-use of sites and buildings through collecting and analysing text sourced from media coverage, government documents and relevant websites. This strategy yielded research material comprising archival and secondary data, and web pages with both text and visual images. Where appropriate, we supplemented this material with promotional documentation collected from the organisations involved at each site. We drew on media coverage to develop narratives of closure histories and debates over re-use. We also used media sources to identify major threads in public discourse about the positive and negative aspects of particular asylum histories and conflicts between public and private interest over development. We recognise that an important characteristic of these sources is that they represent the views of sometimes extremely partial actors, whether individuals or institutions, as expressed at a particular point in time rather than through the often hazy lens of recollection.

We followed standard procedures in working with our largely qualitative research materials. Collectively, we view the collated materials as 'text' that sets out a variety of views of current, re-purposed or derelict asylums. Our aim was to subject this text to a critical reading, thereby developing rich biographies of our chosen case study sites. We read all written materials thoroughly and compared interpretations between ourselves. We used a similar approach with observational records including photographic and cartographic materials, with attention given to interconnections between the written and the visual. From this process we identified key tropes within the materials that constituted common themes that could be placed within the context of wider literatures and theoretical positions. We debated our conclusions initially within our research groups and then with others in seminars and conference presentations.

Conclusion

In Chapter 1 we noted that there has been little research on the fate of the psychiatric asylum. This paucity of previous work carries with it the benefit of being able to craft our own path, both in the development of an approach and the deployment of particular conceptual foundations and investigative methods. There is a down side to this situation of course. Our methodological approach had of necessity to evolve through a process of experimentation, trial and error. We converged on the approach outlined above by recognising the limitations of our initial efforts. The culmination of this process and our final methodological step was to visit all asylum sites in New Zealand to provide a comprehensive, national picture against which to set our interpretation of events and developments at individual sites in all three countries (Kearns et al., 2012).

We see parallels in terms of our engagement with matters theoretical and conceptual. The absence of prior work forced us to adopt a two-part strategy. First, we looked to mature scholarship in cognate and related areas, such as that on memory, heritage, stigma and cultural landscape. From these literatures we drew the inspiration for the development of our notions of strategic forgetting and selective remembrance and our distinction between remembrance and memorialisation. Second, we placed this conceptual repertoire within the context of the relevant policy literature, producing a secure framework in which to locate an understanding of the various national and local policy imperatives guiding and constraining the fortunes of the asylum and the re-use of their sites and buildings. Pulling together the often diverse threads from these background literatures was challenging and inevitably selective, but enabled us to develop the basis for the chapters that follow.

Chapter 3

The Survival of the Idea of Asylum

In this chapter we consider the persistence of the idea of asylum and of its constituent features in an era of community care. Our interest is in the survival of residential mental health care provision in settings that resemble the historic asylum. Specifically, we interrogate the survival of the asylum modality in the private sector, distinguishing this from the persistence of mental health uses on former psychiatric asylum sites in the public sector, which we go on to consider in Chapter 4. We focus on business strategies and on the way that the idea and reality of asylum continues through careful marketing to be 'sold' to a discerning and demanding clientele that is rather different from that associated with the asylum in its heyday. We seek to capture the impact of changing national health policy contexts, the evolving role of ownership arrangements and leadership, and the deployment of therapeutic landscapes and related concepts in sustaining and promoting a form of provisioning that runs counter to the dominant community care modality.

Our approach to the examination of the survival of the psychiatric asylum is based primarily on three case studies of private provisioning. The cases – the Homewood Health Centre (Canada), the Ashburn Clinic (New Zealand) and the mental health facilities of the Priory Group (United Kingdom) – were chosen as representative of different scales of provisioning. Ashburn Hall is considerably smaller than Homewood. Until recently both Ashburn Hall and Homewood were stand-alone facilities (and businesses). The Priory Group, as the name implies, comprises several facilities and mental health provision has always been part of a wider private health care business.

Homewood was founded in 1883 in Guelph, Ontario, Canada. It has 312 beds catering for a wide range of conditions meriting hospitalisation. Other activities include community care, corporate mental health promotion and employee assistance and addiction management programmes. Homewood has a 130-year history as a private-sector provider, and the hospital and related extramural programming are currently operated through the Homewood Health Corporation which, since 2010, has been a wholly-owned subsidiary of Schlegel Health Care. The institution currently known as Ashburn Clinic opened in 1882 in Dunedin, New Zealand. It can accommodate around 100 patients but currently comprises a 65-bed inpatient service and clinics catering for eating disorders, a day programme, a self-care hostel and outpatient services. It is run by the Ashburn Hall Charitable Trust. Ashburn's treatment philosophy is firmly anchored in ideas of therapeutic community which sees patients and staff working together and, where appropriate, sharing in decision making. The Priory Group was founded in 1980 and mental health care provisioning can be traced to the purchase of a single hospital, the Priory, Roehampton. By 2013 the group had 15 facilities offering mental health care services in diverse locations across the UK. Individual Priory hospitals often, though not always, predate the foundation of the company and have their own histories of providing private and/or NHS care. In addition to in-patient psychiatric care, Priory has substantial detoxification services as well as commitments in the areas of executive stress, residential schooling and brain injury.

The remainder of this chapter is organised in three major sections. In the first of these we focus on business strategies as a way of understanding the survival of the private psychiatric asylum in the face of two historical trends – the growing ascendancy of community care as the preferred modality for provision and the increasing primacy of the public sector in both the funding and delivery of care. Given that the Priory Group did not emerge until the early 1980s, our discussion of developments in the 1960s and 1970s draws only on the Homewood (Canada) and Ashburn (New Zealand) cases. We weave the Priory Group into discussions of business strategies in subsequent decades, noting the importance of the changes introduced by Margaret Thatcher's Conservative Government in the 1980s for the rise of private provisioning across the health sector. This observation reminds us that 'business' strategies often involve political as well as business manoeuvres, with strong leadership as the binding force. We conclude this part of the analysis by posing important questions regarding the nature and limitations of private ownership.

In a second section, we examine the marketing of the asylum in the private sector, with a particular focus on the commodification of attributes that align with the concept of 'therapeutic landscape' introduced by Gesler (1992a, b) and discussed in Chapter 2. Drawing on an extensive database developed from web and printed sources in 2004/5 (see Moon et al., 2006), we examine the particular images and metaphors deployed in promotional materials by the case study institutions. We draw out from these case studies evidence for a series of tropes and counter-tropes associated with the deployment in marketing of positive aspects of asylum generally and especially of concepts related to therapeutic landscape. In a third and final section, we reflect on the links between business strategies, business networks and the marketing of asylum. We point to the importance of recent changes in ownership and draw out implications for contemporary and future understanding of the idea of asylum.

Business Strategies: Swimming against Two Tides

The literature on policy shifts in health care provisioning provides some pointers concerning the ability of at least some private-sector residentially-focussed institutions to swim against the twin tides of community care and public provisioning, with one particular theme emerging with respect to the survival of the asylum modality. Seldom deeply buried in the critique of community care was, and is, a popular yearning for residential care, most clearly as a mode of service delivery that, through confinement, could bring about the removal of people with mental health problems from the public gaze (Wolch and Philo, 2000). While this perspective is often about combating dangerousness (Moon, 2000), it nevertheless carries with it important indications of the level of latent public support for residential provision and residual faith in the therapeutic potential of the asylum. Such faith has intermittently been re-kindled in various jurisdictions by reports of the negative consequences of deinstitutionalisation for patients (Joseph and Kearns, 1999) and for communities asked to host community networks of care (Kearns and Joseph, 2000). This residual faith in the asylum model of care has implications for the marketing of residential care as a private sector alternative to the dominant public sector modality of community care, which we will consider later.

If insights into possible reasons for asylum survival are sparse in the general literature on shifts in treatment modalities, those from research on the persistence of private residential mental health care are still rarer. Nonetheless, if the net is cast wide two themes of relevance can be discerned. First, beginning in the 1980s the neo-liberal turn in the broader health policy environment favoured private residential mental health care. A public-private mix in the funding, delivery and regulation of care was acceptable in this policy environment, and the promotion of client choice ensured that people who desired and could afford private residential care would have such care available to them. Even in archetypal publicly-funded and directed national health services, it was evident that the state was far from being a monopolistic provider (Lelliott et al., 1996) and that the mixed economy of care pre-dated the growing attention devoted to it in the 1980s (Mohan, 2002). A second theme concerns the adaptability of private residential mental health care; to change with the times (Bleakley et al., 1991) and to provide flexibility to a public sector needing some residential care but reluctant to take responsibility for its provision (Knapp et al., 1997).

Taken together, the identified themes suggest that the survival of private residential mental health care has been, and still may be, inextricably intertwined with the changing perceptions of public-sector community care. Any transfer of clients from residential care to community care within a broader political and ideological environment in which private ownership and provisioning persists is almost bound to encounter the paradox of choice. Some, particularly those with the necessary financial means, will challenge the paternalist prescription of community care, exercise their own choice, and choose a residential alternative. Care within the 'public city' (Dear, 1980), with its connotations of disadvantage, dispossession, poverty and exclusion might thus be shunned in favour of an elite choice of advantage and seclusion predicated on the ability to afford private residential care. Public demand, a (at least niche) market and a symbiotic relationship with the public sector thus emerge as possible bases for the survival of private residential mental health care.

A generalised understanding of the survival of psychiatric institutions demands consideration of the boundaries between public and private, and asylum and community. It also requires an understanding of apparent movement across these boundaries. To survive, institutions change and adapt over time; moreover they modify the representation of their roles over time. In this way, 'mission movement' shapes the way in which a facility fits itself to changing circumstances.

As noted in Chapter 2, we conceive boundaries between institutional and community care settings and between private and public provisioning as inevitably fuzzy (Gleeson and Kearns, 2001). They do not represent clean breaks. Rather, they are zones through which facilities move in the quest for survival or for greater profitability. Such boundary-crossings occur 'in place': they are contextualised geographically and the specific contingent circumstances of location and national policy are salient in framing institutional survival. Moreover, in an echo of structuration theory (Giddens, 1979; 1984), movement across boundaries is galvanised by agency (the actions of significant individuals or groups) and conditioned by structures (changing policy environments and market conditions). Mission movement across a boundary may result in hybridity. Elements of former institutional arrangements may persist alongside the new, thereby further contesting the dualism inherent in the idea of the boundary. Moreover, individual institutions may confront fuzzy boundaries differentially; for some institutions, ambiguity might create opportunities to be exploited, for others it might presage failure.

Mission Movement

In examining the context and nature of mission movement, we focus primarily on the three decades between 1960 and 1990. These two points in time frame three critical policy shifts: deinstitutionalisation, the expansion of public ownership and provisioning in health care and the rise of neo-liberalism. Prior to deinstitutionalisation, private care was a treatment alternative within the asylum modality. Post-deinstitutionalisation, private asylum care was an exception in a system dominated by publicly funded and delivered care in the community. A second phase, less dramatic in its policy book-ends but critical in terms of bringing into focus possibilities for the re-use of the former psychiatric hospital, extends from the 1990s to the present. It is a phase which we examine in a following section through the lens of marketing.

Readers are directed to Joseph and Moon (2002) and Moon et al. (2006) for a detailed review of mission movement by Homewood and Ashburn Hall. Notably, Homewood developed outpatient services in the early 1950s, otherwise it remained true to its identity as a private residential care facility (Joseph and Moon, 2002). In contrast, Ashburn made more definite moves across the private-public boundary. The first instance of public investment in Ashburn Hall occurred as early as 1910 when the government shared the cost of establishing a neuro-pathological laboratory in Ashburn Hall for medical training (Medlicott, 2001, footnote 3), foreshadowing both future investment by the state and integration into the University of Otago's School of Medicine. A second instance of public investment occurred in 1939 when Dr. Falconer, the medical director, successfully petitioned the government's Mental Hospital Division for the recognition of Ashburn Hall, so that its patients might receive hospital benefits under the Social Security Act 1938. Subsequently, the government usually paid 50 per cent of patient charges (Medlicott, 2001, p. 121). For all intents and purposes, Ashburn had become a hybrid; it remained private in ownership but intimately connected with a public education facility, from which it derived legitimacy and prestige, and with the state as a subsidiser of patient costs. In contrast, Homewood stayed away from the public funding but ventured across the asylum-community boundary by offering outpatient services (Joseph and Moon, 2002). While these differences presaged further divergence in subsequent decades, at the end of the 1950s, other than informal care in the family, there was no practical alternative in either Canada or New Zealand to the asylum. While the public asylum predominated in both countries, few institutions challenged the pre-eminence in the private sector of Homewood and Ashburn respectively.

As noted in Chapter 1, by the early 1960s pressure was building worldwide for the adoption of new treatment modalities involving the deinstitutionalisation of the mentally ill. It is at this critical juncture that the stories of Ashburn and Homewood further diverge. In Canada, the growing critique of the asylum modality coincided with a groundswell of support for the introduction of government-funded universal health care. In contrast, there was little challenge to the then well-established mixed public–private health care system or to the asylum modality in mental health care in New Zealand (Hay, 1988). Canada (and its provinces) took up the challenge of deinstitutionalisation enthusiastically and comprehensively (Williams and Lutterbach, 1976; Joseph and Hall, 1985; Dear and Wolch, 1987). In contrast, the (New Zealand) Mental Health Act 1911, which renamed lunatic asylums as psychiatric hospitals, remained in force until 1969 (Haines and Abbott, 1985). The successor Act was virtually silent about community care, emphasising instead the monitoring of standards of hospital care (Abbot and Kemp, 1994).

As a consequence, by the 1980s mental health care in Canada had changed radically: community had replaced asylum as the preferred site of care and the public sector had become dominant in both the ownership of facilities and the funding of services. In contrast, the New Zealand mental health care system was stalled at the crossroads it had reached over a decade earlier (Hall and Joseph, 1988). Community care initiatives had only tentatively emerged in hospital board settings, and the country's psychiatric hospitals still housed nearly 7,000 people (just under the peak population of 8,261 in 1944) and consumed most of the national mental health care budget (Hall and Joseph, 1988). In concert with the nascent contrasts that had emerged prior to 1960, these distinctive policy contexts shaped the different business strategies of Homewood and Ashburn.

The implementation of universal health care in Canada served to emphasise the anomalous position in Ontario of Homewood as the province's only comprehensive private psychiatric hospital. In 1967, Homewood responded to its atypical status by securing a 'loose arrangement' with the Ontario Ministry of Health for the supply of publicly-funded services to the local community (Homewood Health Centre, 1998). We see this as a pre-eminent example of agency and of the importance of political negotiations in protecting or even enhancing the foundations of a business strategy. It seems that the hospital mobilised arguments based partly on geography: the lack of proximate alternative facilities (Wellington-Dufferin District Health Council, 1996). However, it relied most on its record of achievement and its status as a local and provincial repository of expertise in psychiatric care (Tatham, 1983). A proactive leadership stance was also displayed in terms of meeting new, more stringent, licensing requirements. Homewood was the first psychiatric hospital in Ontario (and only the second in Canada) to be fully accredited by the newly established Canadian Council on Hospital Accreditation.

The securing of access to publicly-insured patients provided a degree of long-term commercial security for Homewood, such that by the late 1990s approximately 60 per cent of the hospital's revenue was derived from the provision of services to the Ontario Ministry of Health (Joseph and Moon, 2002); it was geographically 'grounded' in the local (public) health economy. However, by not accepting capital funding, Homewood retained ownership of its assets and protected its right to admit private patients. Subsequent debates about the future of health care in Canada and its provinces, which have at times engaged actively with ideas of increased private ownership and private delivery of selected services, have assisted Homewood's survival further by making the facility appear less of an anomaly.

In addition to taking a proactive stance with respect to the advent of universal health care, Homewood continued to respond to the challenge of deinstitutionalisation. Despite the fact that the various iterations of the Mental Health Act (see, for example, Government of Ontario, 1998) specifically exempted Homewood from the compulsory provision of community services, it set up an outreach clinic in 1967, financed by transferring public funding equivalent to 42 beds from in-patient services (Tatham, 1983).

In contrast, Ashburn, at least initially, remained largely unchallenged by its policy environment. In the mixed public-private hospital system that had persisted in New Zealand since the 1930s, the status of Ashburn as a private psychiatric hospital was not anomalous. Second, the related tradition of government subsidy of public patients in private hospitals provided a degree of financial security to the hospital that, arguably, sheltered it from the negative consequences of its peripheral geographical position with regard to substantial client markets. Third, the relationship with the University of Otago's School of Medicine provided a buffer between Ashburn and policy changes that elsewhere directly challenged the status of residential care. Thus, unlike Homewood, Ashburn was not forced by a changing policy environment to be proactive and inventive. Indeed, it seems that pressure for change did not build until well into the 1980s. We speculate that the sale of Ashburn Hall by the University of Otago in 1988 was a tipping-point in this process – in part a response to changing financial circumstances in the university sector, but also a response to shifting conceptions as to appropriate modalities of psychiatric care and their teaching in medical schools.

By the mid-1980s, while still under the ownership of the University of Otago, Ashburn Hall was in deep financial difficulty. After breaking even in 1986, it faced another financial loss in 1987. It was reported that, "Ashburn Hall needs more occupants, and more money, if it is to survive" (Cooper, 1988, p. 6). The crisis appears to have had origins in both the low numbers of patients using the facility and the decrease in government subsidies to patients in private hospitals. Until 1979, there had been deep government subsidies. They decreased thereafter, and in 1987 the Department of Health abolished all subsidies to private hospital patients. Ashburn responded by negotiating a deal that saw it exempted from the abolition of the patient subsidy. The subsidy did, however, decrease to cover just 34 per cent of patient costs. At the same time, Ashburn's ability to ensure its future was also hampered by the terms of its Government license; it was only permitted a maximum profit of $NZ 25,000 a year. Any excess funds were required to be put back into reducing patient costs or improving facilities and treatment.

In June 1988, Ashburn's administrators launched a nationwide advertising campaign in major newspapers and magazines to bring attention to " ... the existence, the problems, and the capabilities of the hospital". A number of solutions, short of ultimate closure, were identified, all aimed at achieving a higher occupancy rate. An increase in referrals was seen to be "the best option for financial recovery" (Cooper, 1988, p. 7). In due course, it was an expansion of the suite of services (e.g. relationship difficulties, mood disorders and stress-related illnesses), the Hall's re-branding as Ashburn Clinic under its new owners, the Ashburn Charitable Trust, and a re-assertion of its status as an alternative to the public mental health care system (Ashburn Clinic, 2003) that ensured the survival of the institution. These strategies were essentially the same as those adopted contemporaneously by Homewood (Joseph and Moon, 2002).

The (New Zealand) Mental Health (Compulsory Treatment and Assessment) Act, 1992, engineered a shift of patients with mental health problems from strictly psychiatric hospitals to general hospital and community settings (Joseph and Kearns, 1996). Thus, Ashburn encountered pressures for deinstitutionalisation some 30 years after Homewood. It did so hesitantly, with both the belated implementation of community care and the ongoing organisational reform of the New Zealand health care system creating an atmosphere of endemic uncertainty. This situation was particularly the case for a facility like Ashburn Hall that was, in any case, outside the mainstream of the New Zealand health care system.

Ownership emerged as a particularly important issue for Ashburn, with the period of ownership by the University of Otago arguably problematising its status as a private psychiatric institution. Moreover, the reliance on public underwriting of patient costs created ambiguity for Ashburn because the continuation of public subsidies to patients carried with it controls on profits. For all intents and purposes, this made Ashburn a non-profit institution. In contrast, Homewood kept the state at arm's length and continued to exercise a much greater degree of autonomy, especially in those aspects of its business, such as organisational health, totally outside the scope of public provisioning. Homewood took advantage of changes, beginning in the 1980s and flowering in the 1990s, that we now label an expression of neoliberalism. In the 1990s, for instance, Homewood undertook an internal re-organisation that established a wholly-owned subsidiary, the Homewood Behavioural Health Corporation (Homewood Health Centre, 1998). This sought to provide "behavioural and mental health services on behalf of private sector payers such as employers, health benefit insurers and medical disability management companies" (Homewood Health Corporation, undated, p. 3). In this move, Homewood positioned itself in the lucrative market of corporate health care, while also reducing its dependence on revenue (predominantly from the public purse) from the hospital itself. Subsequent initiatives saw Homewood, in partnership with the Schlegel Corporation, venture into the development of residential care for the elderly in various locations in Ontario.

Looking to the UK case, a broader expression of the extent of privatisation in the health care system, including mental health care, is observable. Beginning in the 1980s, a series of measures implemented by the Thatcher Conservative government made it both feasible and profitable for the private sector to take on new roles within the health care system. By way of example, legislation progressively removed the monopoly of the state and voluntary sectors in the provisioning of residential care for the elderly. Phillips and Vincent (1986) document the extensive conversion of small hotels throughout Britain, but especially in coastal towns, to take advantage of this new business opportunity, which they dubbed the rise of 'petit bourgeois care'. This development involving small 'players' is illustrative of reach of the neoliberal turn in transforming health care delivery through a Gothic horror images permissive approach to the staffing of care settings. Parallels can be seen in the establishment of group homes for psychiatric patients, where the primary emphasis was on meeting the demands of planning legislation rather than health care standards. While subsequent re-regulation of the aforementioned sectors went some way to appease public concerns with the quality of care, the role of small business in health care continued to rise steadily. For example Andrews and Phillips (2005) document the prevalence of small businesses in the provision of complementary health therapies in the UK.

The 1990s witnessed the emergence of the Priory Group as a prominent provider of residential mental health care in the United Kingdom, with the year 2000 a key point in its evolution. In that year, the group, which had been founded in 1980 through the acquisition of the Roehampton site by California-based Community Psychiatric Centres, was acquired by Westminster Health Care. The latter was a prominent provider of nursing homes, and had itself been acquired a year earlier by physician and entrepreneur Chai Patel with the backing of investment bank Goldman Sachs. A few years later, Westminster's nursing homes were sold off and Patel, this time with the backing of venture capital firm Doughty Hanson, engineered a management buy-out of the Priory Group (*The Telegraph*, 10/7/05). The Priory Group was sold on to Dutch bank ABN Ambro in 2005, with the latter itself purchased by The Royal Bank of Scotland (RBS) in 2007. Dr. Patel sold the bulk of his 16 per cent share in the Priory group in 2008 and severed his links with the group. Later in this chapter we will reflect on the implications of these changes in ownership for the Priory Group, which has continued to be the flagship provider of asylum-style care in the UK. Here, we examine the type of marketing conducted by the Priory Group during its period of greatest celebrity under the leadership of the flamboyant Chai Patel. We compare and contrast these strategies with those deployed by Homewood and Ashburn.

Marketing: Deconstructing the Asylum

During the 1980s, successful marketing of a positive image for asylum care became central to the survival and profitability of private providers. The notion of the asylum as a 'removed location' has continued to be attractive (Moon, 2000). Our interest here is with the packaging of removed location and its (re)presentation as a positive selling point to a discerning market. We are particularly interested in the deployment of ideas and images associated with therapeutic landscape as a means of promoting removed location and dealing with negative connotations of institutionalisation.

The Therapeutic Landscape of the Asylum

The asylum appears to be predicated to some degree on ideas of landscapes as therapeutic. Park-like grounds, seclusion and healing through removal from society and exposure to the positive properties of particular places were deeply embodied in traditional notions of asylum as a care delivery modality. In addition, a warm, pleasant atmosphere, in attractive surroundings, was seen as a valid complement to psychotherapy and chemotherapy. Despite this seeming resonance, asylums have not, to our knowledge, been recognised explicitly as therapeutic landscapes in the formal sense. More generally, a conceptual underpinning to the representation of the asylum as a therapeutic landscape is also implicit in research on the inter-relationship of place and health (Jones and Moon, 1992; Kearns, 1993, Kearns and Joseph, 1993; Moon, 1995).

As noted in Chapter 2, the lack of explicit work on the asylum as a therapeutic landscape may, in part, be a consequence of the erasure of the positive therapeutic element in the history of the asylum in favour of a focus on its more recent negative image. This vilification of the asylum in the era of closing asylum places and opening up spaces of care in the community presents a major challenge for the marketing of present-day private asylum provision. We contend that the contemporary construction and presentation of the asylum as a therapeutic landscape is a key response to this challenge in the private sector. Moreover, we suggest that this response is not so much a recovery of the therapeutic past of the historic asylum but more a present-day attempt to commodify some of the components of residential care.

Image-making and the Commodification of the Present-day Asylum

In terms of our focus on present-day asylums, we are interested in both their material reality in the built environment and the ideological presence in the perceptual landscape that this reality generates. Following Kearns et al. (2003), we argue for a recursive link between the material and the ideological. Advertising and promotional texts about present-day asylums can be seen as leading to their legitimation as an ongoing modality of care. However, the casting of such residual provision as a commodity to be desired is dogged by the stigma of the asylum. The historical legacy of confinement, austerity and grim treatments heightens the challenge of promoting and marketing sites of asylum, of successfully commodifying what was historically shunned.

Notwithstanding the challenges from the past, we contend that an important geography is evident in the use of language and image to create position and maintain the place-identity of present-day asylums as therapeutic landscapes. Our interest in this link builds on the work of cultural geographers for whom places and their imaginings are not given, but rather are made (and consumed) through the contested processes of cultural production (e.g., see Gesler and Kearns, 2002). In this respect, places such as hospitals exist not only as empirical entities, but also as social productions, reflecting changing underlying relationships of power, class and cultural expectation. Our analysis also finds roots in the views of those cultural geographers who are increasingly expressing interest in the production and symbolism of urban landscapes (e.g., see Mansvelt, 2005). The selling of an urban lifestyle has become an integral part of an increasingly sophisticated commodification of everyday life, in which images and myths are packaged and (re)presented until they become 'hyperreal' or elevated from the metaphoric into the everyday and taken-for-granted (Holcomb, 1993). Here we are interested, in a sense, in the selling of an 'asylum lifestyle'. To this end, ideas of place marketing in the 'selling cities' literature (e.g., see Bradley et al., 2002; Kearns and Philo, 1993; Madsen, 1992) find parallel in contemporary health care. In this discourse, hospitals, like other parts of the urban realm, become commodified and are rendered attractive to patients and investors through the conscious manipulation of images (e.g., see Kearns and Barnett, 1999).

Marketing the Private Asylum – Tropes and Counter-tropes

Deploying the methods outlined in Chapter 2, we seek to characterise the particular images and metaphors used in promotional materials and to interrogate the underlying discourses that reveal evidence of power and position in the maintenance of place. By discourse we refer to language and writing (printed words as well as inscription in the landscape) intended to " … persuade ourselves and others to a certain way of understanding" (Harvey, 1996, p. 77). It is our contention that the messages about asylum care conveyed in promotional materials may assist in constructing new ways of understanding the very nature of mental health care itself.

As noted earlier, our analysis sought to uncover evidence of therapeutic landscape ideas. We systematically searched for tropes engaging with notions of community, privacy, seclusion and recovery as well as more straightforward themes about landscape, buildings and facilities. Counter-tropes were also sought. These represented textual motifs that challenged the central hypothesis that notions of therapeutic landscape and asylum are integral to the (re)presentation of the case study facilities. Our approach has parallels with studies that have drawn their information from news media and printed institutional marketing documents (Lawrence et al., 2008; Kearns et al., 2003). Initially, we consider each case in turn, beginning with the mental health care hospitals of the Priory Group. According to the Group:

> Priory's aim is to provide the best quality care and services for all and to ensure that these are delivered to the highest standard by professional and committed staff. The Group's values are based on Service, Innovation and Integrity and its purpose is to bring 'Hope, Healing and Sanctuary' to all and to assist each individual to take control of his or her own life within a safe and secure environment (Priory Group, 2005a).

Ideas of safety and separation from a threatening outside world were central to historic notions of asylum, and these qualities remain evident in the presentation of the Priory Hospitals to present-day clients. Two intertwined devices enable the marketing of sanctuary: security and safety. First, there is reference to high quality service delivery. Promotional materials stress the availability of carers providing oversight of the service user and protection from harm. For example: "Exits through the building lead to safe gardens where patients are free to wander under the unobtrusive surveillance of nursing staff" (Priory Group, 2005b). Publicity also emphasises the regular inspection of facilities and the qualifications, networks and accreditation of staff. There is full participation in national inspection schemes and accreditation by the UK Health Quality Service. These claims implicitly draw comparison with other (understaffed, community, public) services. Second, the facilities are promoted as safe and secure in their own right. They are places from which threats and pressures are excluded. Potential disturbance is prevented from entering the user's secluded world. Moreover, this safe, secure world lies within definite boundaries. There is a clear separation between the inside of the asylum world and the outside, everyday world: "It is easy to drive past the Priory in Roehampton with not the faintest idea of what goes on behind its high walls" (Franks, 1998).

Looking inside the asylum world, landscape is a clear theme in the presentation of the Priory Hospitals. As argued above, the objectives of the historic asylum incorporated ideas about landscape as therapy. Asylums sought the promotion of recovery through the calming properties of particular landscapes. The present-day Priory Hospitals explicitly echo this approach: landscape is portrayed as a key element in the promotion of better mental health and ordered, highly designed park-like settings are presented as a characteristic of many of the Priory hospitals in marketing materials. By way of example, the website for the Altrincham Priory Hospital emphasises its rural location and spacious grounds: "situated in its own extensive grounds in the heart of the Cheshire countryside" (Priory Group, 2005c). The website for the Bristol Priory Hospital extends this trope to provide a more explicit link between locational setting and the care of those with mental illness: "Set in four acres of landscaped gardens, The Priory Hospital Bristol offers a tranquil environment for those receiving treatment for psychiatric problems" (Priory Group, 2005d).

Alongside tranquillity and seclusion in these landscape tropes sit clear themes of ownership and privatised exclusion and exclusivity. Field visits substantiated this aspect to the claims made in these promotional materials. Priory's Marchwood site, for example, is accessed via a winding drive through open parkland; the building looks across this vista to an artificial lake and mature woodland. It is invisible from the main road. The more urban Grovelands site benefits from co-location alongside a public park from which it is separated by impressive ornamental fencing. Further themes that are evident in this use of landscape as a promotional device are reassurances about proximity to urban life and references to well-known nearby landscapes as well as simple statements noting the size of the hospitals' grounds. While the latter theme contributes to the motif of seclusion, the former issues offer clients a continuing connectivity with urban life and locate the facilities in areas signifying rural quality: isolation is tempered by accessibility and seclusion takes place in environments of repute.

In effect, what is being offered is the health care equivalent of the country house (hotel) experience. In promoting this opportunity, there is, however, recognition that care and treatment form part of an experience that is about more than landscape. Hints of this are evident in quotes that place the actual hospital buildings in their contexts. Thus the Roehampton Priory Hospital is " ... a most attractive building in a tranquil setting in Roehampton, South London, close to Richmond Park" ((Roehampton) Priory Hospital, 2005). In other cases it is treatment that is linked to landscape. The Priory 'Grange' group of facilities make particular play of this second link. They " ... offer a home and treatment for adults with enduring mental and physical illness. They deliver intensive but highly flexible care programmes in a safe and tranquil setting" (Priory Group, 2005e). The Priory Grange Hospital's Heathfield sites exemplifies this claim: "The space and tranquillity offered by the Unit and its surroundings provide a perfect setting for people who are experiencing severe and enduring mental health problems" (Priory Group, 2005b), as does Ticehurst House: "Founded over 210 years ago, the hospital stands in 48 acres of extensive grounds, providing a calming therapeutic environment for our patients" (Priory Ticehurst House, 2005).

In this last quote, we discern a further theme: appeals to history. These apply, on the one hand, to the provision of care. The Roehampton Priory Hospital, for example, " … first became a hospital in 1866 and is now recognised as one of the foremost private psychiatric hospitals in the United Kingdom" ((Roehampton) Priory Hospital, 2005). The linkage of history and the therapeutic appeal to landscape noted above for Ticehurst House is also extended to Roehampton in a section of the main Priory Group website specifically devoted to the history of the group: "The Priory Group owns two of the oldest private mental health hospitals in the UK: The Priory Hospital Roehampton and The Priory Ticehurst House. The Priory Hospital Roehampton is London's oldest private psychiatric hospital and has been in continuous operation since its launch in 1872, when Dr William Wood moved his patients from Kensington to Roehampton's then country atmosphere, which he felt was conducive to healing" (Priory Group, 2005f).

Appeals to history are also evidenced as architectural signifiers of quality. For the most part, Priory hospitals occupy inherited sites, and thus do not possess many of the innovative purpose-built structures associated with contemporary health care facilities. Historic buildings however clearly form important parts of the therapeutic landscape of the present-day asylum. There is an emphasis on the status of the buildings in publicity material: "The Priory Hospital Hayes Grove is situated close to Hayes Common, just south of Bromley, Kent. Within its spacious grounds the hospital incorporates a listed Queen Anne mansion" (Priory Group, 2005g). At the Chelmsford and Marchwood sites the buildings are reported as Grade II listed and Farm Place is a seventeenth century manor house. The Roehampton hospital is described as " … built in the first part of the 19th century in a style known as Strawberry Hill Gothic" ((Roehampton) Priory Hospital, 2005). It is in fact a 'type-site' for that particular architectural style. Civic responsibility is also implicated in these references to architectural signifiers: "Priory bought Heath House, a Grade II listed building, from the National Health Service in 1991. The House dates back to the 18th century and for eleven years had been empty and almost derelict. After extensive restoration the building was totally renovated back to its original splendour and is now the centre of The Priory Hospital Bristol" (Priory Group, 2005d). By drawing attention to the architectural status of the hospital buildings, we contend that promotional materials implicitly seek to confer a similar aura of status on the users of the buildings. This promotion of quality contributes to the positioning of the Priory Hospitals as high-standard service providers.

The internal configuration and facilities of the hospitals are also of importance in representing asylum to the contemporary public. As the then chief executive of the Priory Group, Dr Chai Patel, has argued: "Our expectations now are very different from what they were in the past when communal bathrooms and shared rooms were taken for granted" (Todd, 2000). To this end, the description of each hospital routinely includes a brief outline of the facilities that a user can expect. At the Priory Hospital Glasgow the linkage of ideas about the therapeutic advantage of the internal environment and the quality of that environment are explicit: "The environment we create for patients is as important as the treatment itself and each patient has the privacy of their own comfortable bedroom with television, telephone and en-suite facilities" (Priory Group, 2005h). At most of the Priory hospitals, hotel-like dining, exercise and social facilities complement these individualised features.

A characteristic of the historic asylum was the idea that patients should have their time occupied with structured activities. Traditionally, the focus was on the outdoors, on agriculture and horticulture. Latterly, elements of sporting activity were introduced. 'Occupational therapy' also came to include significant 'indoor' activity. We can see elements of this trope, updated by a contemporary concern for physical fitness, in the present-day Priory Hospitals. At the Woodburn Priory Hospital "Communal areas provide a pleasant and comfortable environment and patients are encouraged to use facilities for physical exercise both within the hospital grounds and in the locality. There is an art room and activities area. Individual Aromatherapy, Reflexology, group Tai Chi and Yoga sessions are available" (Woodburn Priory Hospital, 2004). While the asylum farm has vanished, horticulture as therapy continues at some hospitals. However, the marketing of the hospitals makes more of the provision of recreational facilities, both within the hospitals and in the form of negotiated exclusive access beyond the hospital walls. The stress on security and safety remains: "Some patients swim at the local hydrotherapy pool, others can take part in carriage driving or ten pin bowling. However, for those who cannot cope with the outside world we offer a safe environment with activities within our boundaries. These include both leisure activities e.g., music, art, computers etc. or therapy such as aromatherapy, physiotherapy, cooking etc." (Priory Group. 2005e).

Visual cues in publicity material provide additional evidence of the use of therapeutic landscapes and the asylum tradition in the presentation of the Priory Hospitals. There is a picture of each hospital on the Priory Group website. These pictures reinforce the points made above about architecture but also touch on the issue of landscape by depicting historic buildings located, in many cases, in park-like settings. By way of illustration, Figure 3.1 shows the historic buildings of the North London Priory Hospital.

Turning to the Ashburn Clinic, publicity signals that it caters to four types of patient. These are people, who "fail the state system; who need psychotherapeutic treatment; who need a longer term residential environment; or who are attracted to our setting rather than the public alternative" (Ashburn Clinic, undated a, p. 2). In contrast to the Priory Hospitals where the tropes are implicit, notions of asylum and therapeutic landscape are explicit in the present-day objectives and presentation of Ashburn: "Still today the design of the hospital and the grounds are integral to providing a therapeutic environment distinctly different to most psychiatric institutions" (Ashburn Clinic undated a, p. 2). This quote is important in that, not only does it engage directly with the central concerns of this chapter, but it also clearly positions Ashburn as different and as an institution. It is different in that it *is* an institution and different in that it *is* avowedly an institution in the era of community care.

Figure 3.1   The North London Priory Hospital, Southgate, UK

In a lengthy passage on the role of Ashburn Clinic, an information booklet specifically describes the hospital's presentation as an asylum:

> The importance of the meaning of the word asylum for some patients has recently been affirmed in the literature. The Ashburn Clinic has always had a role, with some local patients, in providing asylum. In past decades people who were chronically psychiatrically compromised were encouraged to stay in institutions in the belief that this afforded them a better quality of life. For some this is still true, and for many older patients leaving hospital is now not a real possibility. The Ashburn Clinic, therefore, continues to have a role in the provision of asylum ... (Ashburn Clinic, undated a, p. 2).

Here we see a presentation that alludes to expert opinion (the literature) and to historical continuity (" ... has always", " ... past decades"). A service need is identified and Ashburn is presented as the solution to that need.

   Landscape is a major trope in the representation of Ashburn Clinic. Though in a classic urban fringe location, it is presented, like certain of the Priory Hospitals, as close to urban life. Greatest emphasis is however on the peace, tranquillity, diversity and extent of the Clinic's grounds:

> The Ashburn Clinic is situated 10 minutes from the city centre and is surrounded by farmland. The extensive grounds are filled with flowers, shrubs and exotic trees, which attract a variety of native birds. A serene setting for just sitting or strolling, with large lawns for outdoor sports (Ashburn Clinic, undated b, p. 2).

The reference to serenity is significant. It links back to the therapeutic role of the landscape. It also links Ashburn to 'New Age' notions of spiritual rebirth. This viewpoint is echoed in recurrent visual images on the Ashburn website and in current newsletters and guides for intending users. Images of greenery, exotic planting and blossom are used to present Ashburn as a garden retreat. Indeed, the frontispiece banner on Ashburn's website stated:

> The Ashburn Clinic gardens are a place of quiet reflection, where the cycles of nature are a reminder of the regrowth that is fostered here. (Ashburn Clinic, 2005)

As with the Priory Hospitals, Ashburn valorises the integration of its buildings with its physical landscape. It is also made clear that the built environment is well-maintained and high quality: "Our totally refurbished accommodation is beautifully set in several acres of lawn, trees and rhododendrons" (Ashburn Clinic, 2005). This therapeutic articulation of the physical and the built environment is most clearly articulated by the recent development of Te Whare Mahana o nga hau e wha (the warm house of the four winds): "In an old orchard beside a stream we have created a quiet spiritual place for people to sit and reflect" (Ashburn Clinic, undated a, p. 2) This building, designed by a leading New Zealand architectural partnership, seeks to capture both traditional (western) notions of asylum and therapeutic healing and Māori concepts of spiritual well-being and recovery (Figure 3.2). This is a bold initiative that echoes the bicultural spirit of developments at the former (public) Tokanui Hospital, which closed in the late 1990s (Joseph and Kearns, 1996; 1999).

The interior facilities of Ashburn receive rather less attention in publicity materials compared to those of the Priory Hospitals. Users are encouraged to personalise their rooms and there is some description of communal areas, although provision appears rather more Spartan than that at the Priory hospitals. This distinction can be traced to an important difference between the two operations. While Priory emphasises choice, the quality of interior facilities and a hotel-like approach, Ashburn presents itself as a therapeutic community where even the welcome in publicity is from 'clinic staff and patients'. We will return to this matter in our discussion of counter-tropes but, for the moment, it is relevant to note that, "In living together and performing the necessary domestic and administrative tasks, a sense of belonging, safety and responsibility grows" (Ashburn Clinic, undated b, p. 5). The users of Ashburn play a part in running the Clinic. This is part of the recovery regime; it is a form of occupational therapy but it also means that there is a more workaday image with less emphasis on services provided by others and the opulence of the surroundings.

**Figure 3.2   Ashburn, Dunedin, New Zealand: The House of the
Four Winds**

Nonetheless, structured recreational activity is an important theme at Ashburn. The lawns provide a setting for outdoor activity while, "Alongside our main building is our recreation hall which is equipped for a variety of indoor games including volleyball, badminton, netball and indoor bowls. Next to this is an outdoor tennis court" (Ashburn Clinic, undated a, p. 2) A tour of the premises also reveals a large and well-loved billiard room. Again, we see the concern to promote sporting activity rather than the farming or horticultural activities that historically characterised the asylum. There are however notable differences between Ashburn and the Priory Hospitals. The activities at Ashburn tend to be team sports and they tend to require greater levels of physical exertion. While it is tempting to view this contrast as a reflection of difference in clientele or even gross stereotypes of national cultures (New Zealand vs UK), it is perhaps best seen as another manifestation of the distinction between a therapeutic community where people work together for healing and a hospital chain that prides itself on offering high quality choices in a quasi-hotel environment. We now add to this mix the images promoted by Homewood (Figure 3.3).

Figure 3.3   Homewood, Guelph, Canada: The Grounds

> From the beginning, Homewood has emphasized the importance of a therapeutic
> environment for healing the mind, body and spirit. Second growth forests,
> the meandering Speed River, maintained lawns, grand vistas and historically
> significant architecture create an environment that is tranquil, serene and
> reminiscent of a bygone era (Homewood Health Centre, 2003a).

Notwithstanding this explicit comment about its grounds, the Homewood Health
Centre provides rather less evidence on its website and in its publicity material of
notions of therapeutic environments or continuity with historic ideas of asylum.
Nevertheless, a range of relevant tropes are present that resonate with the themes
identified above: "Homewood Health Centre is a leader in mental health and
addiction treatment, providing specialized psychiatric services to all Canadians.
Located in Guelph, Ontario, in a beautiful setting on the banks of the Speed River,
Homewood has been improving lives since 1883" (Homewood Health Centre,
2003b). Here we again see claims to excellence, historical referencing and place.

A user testimonial, one of a series that together constitute a novel feature of
the presentation of Homewood, provides an indication that the hospital offers its
residents secure safe sanctuary. Indeed, the testimonial makes an implicit contrast
between the negative image of the (historic) asylum and its more positive aspects:
"I had my fears about where I was going. No-one had really told me about what
goes on in these places. I feared the worst. But without question it was the best

move I've made in my life. The hospital environment provides protection from the sources of stress. In a short period of time you begin to calm down ..." (Homewood Health Centre, 2003c),

Landscape and architecture interact in many of the images on the Homewood website. One view is shown of the main hospital building from the grounds; the impression is one of order, calm and authority. Other images concern the provision of different therapies. Two sets are particularly relevant for the issues raised in this chapter. First, the presentation of pastoral therapy echoes the notions raised by Te Whare Mahana o nga hau e wha at the Ashburn Clinic. Homewood boasts a labyrinth where spiritual health is grounded in place-specific activity. Second, Homewood makes active use of its grounds: "Homewood Health Centre hosts the largest and longest-running horticultural therapy program in Canada. Horticultural therapy promotes a 'natural' sense of wellness, and is an adjunctive therapy in all treatment programs offered at Homewood. ... Therapy takes place in the newly constructed 'state-of-the-art' conservatory and classroom, as well as on 47 acres of garden and wooded area, patio and container gardens, and raised garden beds" (Homewood Health Centre, 2003c). Opportunities for horticultural therapy are evident throughout the site (see Figure 3.4).

Figure 3.4   Homewood, Guelph, Canada: Horticultural Opportunities

Where the Priory Hospitals chose to present their indoor landscapes as akin to those of a hotel and Ashburn emphasised the therapeutic community dimension, Homewood, at the time of our visits, exemplified a third approach. Space was not necessarily open to personalisation. Accommodation was available in private rooms but also, in the medicalised hospital tradition, in shared rooms and on wards. There were visiting hours and no telephones in rooms; televisions were located in patient lounges. Dining was recognised as a therapeutic opportunity but also presented in terms of nutritional requirements. Meal times were assigned. Overall, the impression is that, rather more than the other case study facilities, Homewood presents itself as a relatively traditional hospital.

So far we have carefully reviewed representations of our case study facilities, seeking evidence for the deployment of tropes associated with therapeutic landscapes and positive aspects of asylum. Our contention has been that these tropes are central to the representation of our case study facilities in the present-day era of community care; they both distinguish and valorise this rather different form of care provision. To assure this analysis, we also need to consider counter-tropes: themes evident in our source material that suggest representations other than those on which we have concentrated thus far.

In the case of the Priory hospitals, one such counter-trope is provided by the theme of choice. Of course, the package of high quality asylum care in a therapeutic setting that we have discussed in previous sections is a central part of constructing a commodity that a discerning public will choose to purchase. Our research material does however give some emphasis to choice as a distinct construct in its own right. For example: "To complement our care for the mind, we ensure that the choice, preparation, presentation, and quality of food is appealing to patients. At each meal there is a choice of hot and cold dishes to suit different tastes and nutritional needs. We also cater for special dietary requirement" (Woodbourne Priory Hospital, 2004). This is nothing to do with the selling ideas about asylum or therapeutic landscapes; it is simply about presenting a quality service. This emphasis on quality service is complemented in the media by numerous references to the celebrities drawn from the worlds of music, television and elsewhere who have sought out the services of a Priory Hospital. Such a choice, by those who can afford the best, is both an endorsement of quality and a transfer of celebrity.

Our research material on the Ashburn Clinic reveals a strong commitment to the philosophy of the therapeutic community. Though we have drawn parallels with the theme of therapeutic landscape, there is some case for seeing this idea as a counter-trope. While therapeutic landscapes are in our view associated with the healing properties of the physical or built environment, therapeutic communities operate on a more sociological level in which place-as-setting may matter less than place-as-group-dynamic. Thus, patients live together with carers forming relationships which " ... provide the human warmth, support and understanding that is necessary for healing" and " ... treatment at the Ashburn Clinic revolves around participation in community activities, meetings, and interactions in community living. Our emphasis is on working with the individual within the therapeutic community using a psychodynamic approach to promote individual and personal growth" (Ashburn Clinic, undated b). Though these processes certainly unfold *in* a particular place isolated from the surrounding society, the extent to which it is additionally crucial to represent that place as a therapeutic landscape or a benign shadow of the historic asylum is, in a sense, immaterial. There is some evidence of this counter-trope of therapeutic community at Homewood. The key counter-trope evident in the research material on Homewood is however continuity of care. Though the Priory Hospitals provide a range of services, their key focus is on residential care. Homewood, in contrast, is also active in community psychiatry.

On the weight of evidence, a measured conclusion from our research material would be that asylum and therapeutic landscape are significant general tropes in the representation of contemporary institutionally-based mental care in the private sector. However, their relative significance for the three case studies examined in the present chapter varies, and in all three cases there are important counter tropes. The concepts of asylum and therapeutic landscape are least significant for Homewood where there is a countervailing stress on a medical model of care and integration with community care. In contrast, both concepts are, in different ways, important to the presentation of Ashburn and the Priory Hospitals. In the former this importance arises from an articulation with therapeutic community; in the latter it reflects market positioning as a quality service.

Reflections on the Private Psychiatric Asylum

Priory, Ashburn and Homewood have few peers in their national contexts. For much of the post-deinstitutionalisation era they have only had to compete with what they are not: psychiatric hospital wards or community care provision. They were anomalies: hospitals in a post-asylum era and (at least in part) fee-for-service facilities in state-funded health systems. What was not anomalous in terms of the health sector in general was the role these private facilities occupied within the suite of possibilities confronting those in need of mental health care. Their predecessors, the traditional public-sector asylums, had come to represent uniformity, standardisation and scale economy for the population at large, but not necessarily for those who could afford the cost of private care. There was a preference in the public sector for very large-scale facilities run on regimented lines serving well-delineated geographical populations. Within this fixation on delivering a basic service at low cost, the core concept of asylum as a place of sanctuary was almost invariably marginalised or even lost.

With these ideas in mind, it seems apposite to term the persistence of private residential mental health care a postmodernist phenomenon and to view our case studies as sites of resistance to the dominant modality of community care. Through their survival they represent and proclaim the viability of a past modality of care, albeit an idealised memory of that modality. In this interpretation the private sector asylum is meeting a societal need for which the state has eschewed responsibility.

Residual demand for residential care ensured that this policy space could be filled at an acceptable financial rate of return. The contemporary survival of the asylum in the private sector could thus be seen as a market-led response to opportunity and demand; its flexibility and niche activity viewed as the antithesis of a modernist project of hegemonic community care. Place, in the form of a repackaging of traditional notions of asylum, clearly plays a role in this flexibility and in ensuring an effective fit with consumer demand.

The key challenge for the present-day private asylum is assuredly that of image. Promoting (private) asylum facilities must involve countering the legacy of the poor reputation of the historic asylum. There must also be efforts to persuade prospective patients, their families or third-party payers of the value of expensive, potentially long-term and often distant residential care. Both Ashburn and Homewood originally developed as alternatives to poorly-reputed public asylums; their continued existence, and that of the Priory Hospitals, owes at least something to a market seeking to avoid the downside of community care. The historical weight of these ambivalent assessments provides an opportunity for a critical perspective on the present-day retention and reinvention of the asylum. Indeed, the (albeit limited) rise of private psychiatric hospitals can be seen as part of a broader restructuring of health care that is opening space for both private capital and privatised need.

Within the contemporary landscape of mental health care, private asylums have a visibility that is markedly disproportionate to their size. This is testimony to the enthusiasm with which leadership of the institutions has worked to re-sculpt the image of residential care. In Chapter 4, we will consider survival and re-emergence stories in the public sector – of psychiatric asylums or of psychiatric services on former asylum sites. We will argue that such survivals and re-emergences are more numerous than is commonly understood. In large part, this lack of visibility is deliberate and accomplished through re-naming and other strategies (see Chapter 2). The degree of success in promoting poor visibility is such that the public gaze is re-directed to private sector facilities, even when these are smaller and arguably of less significance in the landscape of care. Thus, in New Zealand, where all former asylums have officially been closed, residential facilities of significant scale persist on several sites. For example, in Chapters 4 and 5 we consider the Mason Clinic on the site of the former Carrington/Oakley Hospital (Auckland). Parallels in the United Kingdom can be found in the secure unit at Knowle, near Southampton (see Chapters 4 and 6). In Ontario, Canada, residential care has re-emerged on several former asylum sites, with yet others planned (see Chapter 4).

What does the future hold for the private psychiatric asylum and for its role in maintaining, at least symbolically, the idea of asylum? Clues can be found in the earlier discussions of business strategies and marketing. With respect to business strategies, the survival of private asylums has come to rely increasingly upon the support of the public sector. For both Homewood (allocation of publicly funded beds secured in the 1960s) and Ashburn (access to patient subsidies from the

1930s), survival would arguably have been impossible without access to public funding to support operations. This symbiotic relationship with the public sector (and purse) is replicated in the operations of the Priory Group, with 85 per cent of its services delivered in partnership with the NHS (Priory Group website). Legislation enables this relationship and it is further underpinned by close links with both business and professional networks.

While the public sector may be willing to continue its quiet support of a 'private' residential option, there are threats, both external and internal, to the status quo. Externally, the re-emergence of residential care (regardless of what it is called) in the public sector may squeeze out private providers competing for the same patient funding. Internally, and probably more significant in the short term, even with a continued flow of public funds the business survival of private asylums may be affected by ownership change and associated mission movement. We have already alluded to the case of Ashburn, where a close partnership with the University of Otago developed over several decades resulted in a period of ownership by the university in the 1980s, amounting effectively to a period of withdrawal from the private sector. It was the decision by the university to divest itself of Ashburn that precipitated a crisis that very nearly resulted in closure. There are some parallels in the recent history of Homewood. Here, the Homewood Health Corporation entered into a series of partnerships with the Schlegel Corporation, a large-scale provider of long-term care and seniors' retirement home facilities, such that in 2010 the latter acquired majority control of the Corporation (Homewood Website). Following a period of transition, the new CEO of Homewood, Jagoda Pike, is the first non-medical professional to hold this position. The future of the psychiatric asylum component of Homewood will now be arbitrated within a much larger and more diverse business entity. This said, the issues raised by the sale of Homewood pale in comparison with those brought into focus by sequential changes in ownership of the Priory Group in recent years.

As noted earlier, in 2005 the Priory Group was sold to the Dutch bank ABN Amro by its owners Doughty Hanson, a venture capital company. Ownership passed to the Royal Bank of Scotland (RBS) when RBS purchased ABN Amro in 2007. Massive profits were made by Doughty Hanson and by Chai Patel, CEO and major shareholder, through this sale (*Financial Times*, 31/3/08). A government bail-out in the face of the imminent collapse of RBS during the 2008 financial crises resulted in effective nationalisation (and, indirectly, state ownership of the Priory Group) (Companies and Markets.com). In 2011, RBS sold the Priory Group to Advent International, an American private equity group with extensive international healthcare holdings. In announcing the purchase, Advent International declared that it aimed to identify "opportunities for consolidation that emerge in a sector driven by a strong combination of social trends and economic and regulatory reform" (*Financial Times*, 19/1/2011). The same article went on to note that "the healthcare industry is facing a tough couple of years, amid pressure on local authorities and the National Health Service to reduce the fees they pay private sector operators".

Thus we see the private asylum as caught between corporate owners who demand returns on their investment and public sector funders increasingly reluctant to transfer funds from their shrinking budgets to underwrite such profits. The response to such pressures may be renewed efforts in marketing, directed towards the potential consumers (patients and their families) and toward the public health care funding authorities and insurance companies. There may also be a withdrawal from more expensive forms of residential care as owners seek to focus their investment on profitable aspects of the health care 'industry' and to eliminate underperforming or chronically expensive 'product lines'. In so doing, we observe corporate interests increasingly in control of private asylums and, through their actions we see the progressive narrowing of the very notion of the asylum. The wider survival of notions of asylum in the private sector remains precariously dependent on public funding. In this sense, the boundary between public and private has become less distinct and more porous. To date, this has, on balance, been to the benefit of private providers. However, with the re-emergence of a re-branded residential option in the public sector, private providers may find themselves increasingly reliant on 'true' private patients (as opposed to state funded patients) and increasingly drawn to areas of activity eschewed by the state, including the interface of mental health care and care of the elderly.

Chapter 4

On-site Survival: The Contemporary Practice of Mental Health Care on Former Asylum Sites

This chapter considers the persistence of mental health care activities on former asylum sites. This persistence can take many forms, the most obvious being the continued use of former asylum buildings for residential mental health care, for the delivery of out-patient services or for services targeted at particular client groups. It may involve the continuation in situ of a pre-existing service. Alternatively, services may be re-located from elsewhere, either into existing buildings or into new, custom-built facilities. Transfer of services from former asylum buildings into new custom-built facilities can also occur within the confines of a particular site. As a collective result of some or all of the above, a sub-set of former asylum sites remain wholly dedicated to mental health care and may offer a range of services reminiscent of the former psychiatric asylum itself. Others retain mental health functions on only part of the site while some remain involved only in the administration of mental health care. In a small number of somewhat specific cases, involvement with mental health is maintained not through the administration or delivery of care but through the memorialisation of its past delivery in on-site museums.

We devote the majority of the chapter to an examination of instances of *complete or partial retention,* paying particular attention to their systemic and idiosyncratic origins. We argue that, while far from normative in its unfolding, retention can almost invariably be linked to 'policy' in two ways. First, retention has been facilitated by the difficulty of implementing the transition to community care and, second, it has been encouraged by the ongoing and pervasive impact of a suite of controls related to the regulation of land use and the health care estate. We believe that an understanding of this context is in many senses foundational to the appreciation of the forces affecting re-use of former asylum sites and buildings, especially in terms of the ongoing impacts of stigma.

In the next section we examine the major forces promoting partial or full retention of mental health care services and consider the highly differentiated international patterns of survival. This review provides the context for the introduction of two contrasting examples of retention – a single former asylum in the UK (St James' Hospital, Portsmouth) and a network of former asylums in the southern portion of Ontario, Canada. In a following section, we then highlight two specific forms of on-site survival that carry with them very particular implications for the remembrance of the former asylum. First, we examine the continued presence of 'secure' or 'forensic' services, accommodating people at the interface of health care and criminal justice systems. These are a relatively common (and highly stigmatised) form of care that are often retained on former asylum sites as well as being a form of provision that has persisted on sites reserved specifically for such use. We introduce the Mason Clinic on the former Carrington Hospital site in Auckland (New Zealand) as an exemplar, noting that in this instance this very distinctive form of partial survival has occurred alongside the conversion of most of the former asylum site and its buildings into a tertiary education facility (see Chapter 5). Second, we examine the commemoration of the asylum era on sites of former or continuing use through museums, monuments and displays. Porirua (New Zealand) and Glenside (UK) are offered as examples, though we also note others with varying scales of activity. In a concluding section, we connect back to the theme of survival by noting that the continuation of mental health services on former asylum sites can itself constitute a living memorial to the heritage of asylum care.

Complete or Partial Retention

The decision to move away from asylum-based care towards provision within community settings can certainly be seen as significant in health care policy, but as noted in Chapter 2, it would be incorrect to see it as a single point in time or even perhaps as a single decision. It was, like most policies, the outcome of a process of deliberation that took place over a period of years. The speed of this process and the prevarications, delays and accommodations that accompanied its final outcome varied. National variations were inevitably accompanied by a spatial dimension within jurisdictions. As national governments moved to close asylums and open up new spaces of care in the community, they found that, at the local level, those responsible for mental health varied in their effectiveness at securing alternative provision.

This was certainly the case in the UK (Craig and Timms, 1992). A similarly differentiated closure process, replete with spatial and temporal lags, was also evident in our other case study countries. In New Zealand, itself arguably a laggard country as far as the move to community care was concerned, the closure process took some 15 years (Kearns et al., 2012). In Canada, closures were earlier and swifter but far from synchronous (see Chapter 1). Elsewhere, the situation was similar. Coldefy (2012) reviews the progress of deinstitutionalisation in Italy, Germany and France as well as the UK. In Italy, a legislated change of direction in the late 1960s was progressing only slowly a decade later, prompting radical administrative reform, the closure of all asylum facilities and a 68 per cent reduction in residential mental health beds. In France and in Germany a more mixed approach was adopted. Both countries developed community-based facilities alongside a continuing role for relatively large-scale hospitals within care networks. Nonetheless, bed numbers in psychiatric hospitals reduced by half in France and by rather more in Germany where reunification makes long-term changes difficult to assess with complete confidence.

It is possible to draw three important threads from this picture. First, variations in the speed of the closure process meant that large-scale asylum facilities could linger in situations where political commitment or local circumstances did not result in a rapid movement to community care. Second, even when change was possible, there was widely perceived to be a continuing need for hospital provision for certain types of service user. Indeed, there has been an international consensus over who such users might be: acute admissions to stabilise conditions and people on the interface between the mental health and criminal justice systems. We examine the latter group in more detail later in this chapter in connection with the survival or siting of forensic units on former asylum sites, but for the moment it is necessary to note that there has been less consensus over the appropriate setting for acute mental health care. At issue is how long the period of residential (hospital) care should be and whether it should be in a setting exclusively devoted to mental health care or in a general hospital setting. In essence it is this distinction that lies behind the difference between France, Germany and the arguably stereotypical situation in the UK where, by 1971, the policy objective had become the eventual transfer of all in-patient mental health services to general hospitals. Third, it was acknowledged that certain long-stay service users were difficult to place in truly community settings as a consequence of their diagnosis or social factors, including the impact of previous institutionalisation (Trieman and Leff, 1998). For this last group, small-scale staffed residential facilities were needed. While new build or conversions in community settings were a frequent response, opportunities were also available in villas and outbuildings on former asylum sites. The latter were often an attractive option in terms of cost and, especially, the avoidance of complications arising from potential community opposition to the siting of new facilities. In situations where available asylum sites lay within built-up urban areas, administrators in the health sector might even imagine services placed on such sites to be 'community located'. We will note such an instance later in the case of St. James' Hospital, Portsmouth.

The inertia accompanying policy implementation was paralleled by difficulties encountered in the constituent processes of change. The move to community-based provision was not simply a policy decision that could proceed once local plans for re-provisioning were in place. Such local plans needed to be resourced and converted from plan to reality. As noted in Chapter 2, stigmatised though asylum sites undoubtedly were, the root of that stigma lay in societal attitudes to mental health problems. Such stigma transcended modalities of care provision and cast a shadow over ambitions for care in the community. The considerable literature that developed on local opposition to community-based mental health care charts the difficulties associated with the move away from the asylum (Dear and Taylor, 1982; Smith and Giggs, 1988; Wolch and Philo, 2000). It was clear that society was not necessarily ready to accommodate people with mental health problems in everyday settings and viewed such provision as 'noxious', on a par with prisons, waste dumps and polluting industry (Dear et al., 1977; Smith and Hanham, 1981; Takahashi and Dear, 1997). Subsequent well-publicised failures of community care (Moon, 2000) saw a revalorisation of containment that matched the nascent public support for the principle of asylum identified in Chapter 3. This societal opposition to the establishment of alternative provision arguably prolonged the life of asylums in situations where care managers were able to delay deinstitutionalisation until safe and appropriate community provision could be made available.

Delays in the run-down of asylums could also eventuate for economic rather than societal reasons. Though the escalating costs of maintaining the asylum system with its ageing and often outdated buildings may have been one reason for the turn to community care, the costs of community care were rapidly recognised as not inconsiderable in their own right (Goodwin, 1997; Jones, 1993; Scull, 1984). Adding the costs of maintaining asylum buildings during an often-prolonged closure phase to the costs of developing new community facilities inevitably increased the overall cost of deinstitutionalisation. Moreover, these costs were affected by both macro-economic and organisational contexts. In the UK and New Zealand periods of asylum closure coincided with downturns in the economy and constraints on health care spending that arguably led to both under-investment in community care and to the continued operation of poorly maintained asylum facilities. In periods of economic prosperity, the ability to dispose of former asylums was dependent on property markets and the local supply of, and demand for, development land. Joseph et al. (2009) show how the relative remoteness of many asylums in New Zealand could stymie attempts to dispose of sites even in times of prosperity. Generally, however, economic prosperity tended to be a signal for swifter disposal. Stigmatised, run-down but historic infrastructure could be re-invented as heritage buildings with accompanying parkland; stigma could be transcended by economic opportunity. Examples abound in the UK, particularly on the outskirts of London where luxury gated communities were developed on several sites, most notably the Manor, Horton and Long Grove Hospitals in the Epsom Cluster, the Royal Holloway Sanatorium and Colney Hatch Asylum (see Chapter 6 for additional examples).

In the UK, sequential reorganisations of the health and social care system also ensured that responsibility for closure and re-provisioning became a complex process involving numerous organisations and consultation requirements. Our case study of St James' Hospital, Portsmouth, later in this chapter exemplifies how this complexity continues to frame the fate of one of the few former asylums that still deliver mental health care. In addition, reorganisations required managing, and this provided another impetus for continuing health care use – not for care delivery but for its management.

For social, economic and organisational reasons and through the inertia of policy, asylums could therefore linger on in whole or partial (mental) health care use, either for direct care provision or as management offices, into the era of community care. Continued survival to the present day is, in many ways, a serendipitous locally-contingent outcome of particular combinations of these processes. Generally speaking, it exhibits only a crude spatiality. In the UK, land demands and population pressures have ensured that there are few surviving asylums in South-east England while examples are more numerous in Wales. In Ontario, prior land ownership has enabled something of a rebirth of the residential modality of care on selected sites as the provincial government has maintained its control of the estates of former asylums. Historically, the province was reluctant to dispose of these estates, in large part because it was difficult to find alternative sites that would not generate public opposition to sites of mental health care provision. As will be discussed in more detail below, re-providing large-scale residential mental health care on such sites has exploited this land bank and been advantageous from a planning perspective.

Looking within asylum sites, it is possible to discern a more overt geography to (partial) survival. The disposal and re-use of parts of asylum sites is not without a certain spatial logic reminiscent of classic models of city structure. Areas of asylum sites that adjoin residential areas or have a clear locational advantage in terms of their connectivity are likely to be more attractive on the land market and more amenable to sale. Communities will support access to parkland on asylum grounds; and new developments will favour those parts of asylum sites that are close to road networks or can otherwise most easily be incorporated into adjoining developments. Retained (or renovated) mental health uses may similarly be connected or, alternatively, excluded: remaindered on the more obscure edge of the site. Connected elements of retained health care will, following the literature on 'noxious facilities', be more likely to be administrative services or less contentious services, perhaps involving children or older people. Exclusionary retention in contrast would be more likely for forensic services conforming to the public equation of dangerousness with mental ill-health. On occasion, however, the margins or edges of sites can be preferred for services, notably in situations where a location confers a semi-rural therapeutic landscape. Wherever mental health service uses persist on a site, there may also be evidence of *cordons sanitaires* preserving a separation between mental health and other uses. This may take the form of parkland, walls or fences, or even car parks. These boundary zones can, of course, be eroded over time as land disposal runs its course: for example, land transfers between the state government and Eastern Michigan University have now effectively erased the boundary between the University and the residual building of the Kalamazoo State Hospital, US (see Chapter 5), conversely residential development at Sunnyside, Christchurch (New Zealand), has seen the erection of a barrier separating it from the former asylum grounds occupied by the rebranded Hillmorton Hospital (see Chapter 6).

The notion of rebranding is the final theme in this section. Mental health services may persist at former asylum locations, across the whole or part of the site, but this persistence may not always be immediately evident. As discussed in

Chapter 2, identifying continuing use can require considerable detective work. In some cases location may obscure continuity, with surviving uses hidden away on more remote, less accessible or less visible parts of a site. More often, however, an ongoing mental health use may be obscured linguistically: facilities are named (or re-named) in ways that do not immediately point to their use for mental health care. A former asylum may continue to host a smaller scale mental health facility with an entirely different name; on occasion such a facility may even occupy new buildings. Thus mental health care continues on the Porirua site in Wellington (New Zealand) but does not operate in the old Porirua buildings or use the Porirua name. The service reorganisations in the UK noted above contribute significantly to the complexity of rebranding on sites split between different successor bodies with their own predilections for naming or re-naming services.

Rebranding through renaming manifests what Berg and Kearns (1996) have termed 'norming': a strategic attempt to create symbolic distance from a difficult history. As mental health care has shifted into the community and away from the asylum, the movement has been accompanied by a turn to facility names that similarly embrace 'normality'. Allusions to health care disappear in favour of names that evoke peace, tranquillity and home. This change mirrors past naming trajectories. Throughout the 20th century there was a move away from the use of the term asylum towards a medicalised deployment of the name 'hospital'. This terminology lingers on retained sites whether in the public or the private sector. Perhaps the most powerful aspect of this rebranding, however, is the naming of the client group served by the facility. It is certainly no longer the case that a continuing mental health use can be identified by a facility labelled as an asylum, much less an asylum for the insane. Clues about retained use may point to mental health, perhaps to psychiatric care or to a particular client group, but are more often simply metaphorical allusions. The long shadow of past site-specific stigma is so powerful that reference to past identities is generally erased. We now explore this and similar processes in more depth in two case studies.

Retention and the County Asylum System in England and Wales – St James' Hospital, Portsmouth

The popular archetype of the psychiatric asylum as a forbidding enclosed world of monumental buildings contained behind encircling walls is perhaps best captured by two 'asylum systems'. One is the US State Hospital system (Dowdall, 1996; Yanni, 2007); the other is the County Asylum system of England and Wales (Melling and Forsythe, 2006). We will examine the latter in our first case study of retention. We begin with an overview of the system and proceed to an examination of one facility, St James' Hospital Portsmouth, where mental health care uses have persisted into the present.

The County Asylum system had its origins in the changing social environment of the early 19th century. Philo (1987a,b, 2004) has provided extensive analyses of the balance between public and private provision and the related emergence of systematised public care alongside prototypical care businesses catering to a more affluent clientele. At risk of over-simplification, the development of 'moral' therapeutic regimes at the end of the 18th century at model facilities such as the York Retreat pointed the way to new forms of care. With its stress on therapeutic landscapes and recovery through activity, the new moral care was rapidly found to be an effective as well as humane alternative to previous modes of treatment. Initial provision was private or charitable but the burgeoning population growth that accompanied the industrial revolution ensured that an increasing demand came from poorer people. The scene was set for government intervention. The County Asylums Act 1808 gave voluntary powers to county councils to build asylums funded by local taxes to house 'pauper lunatics'. Nine were built initially, beginning in Nottinghamshire in 1811 and followed by Bedford (1812). A further 20 were subsequently built. These extended the principles of moral treatment to the poor, but the voluntary nature of the 1808 Act ensured that the geographies of demand and supply did not always match. This situation was remedied by a second County Asylums Act in 1845 that formalised the medicalisation of mental health and legally obliged counties to ensure provision. Over 100 asylums were built following the 1845 Act, variously titled as county or borough asylums depending on the jurisdiction funding construction and providing clientele. A further Act in 1890 formally opened the asylum system to the non-pauper population.

The majority of County Asylums were built to standards commensurate with Victorian local pride and contemporary best practice in mental health care. As Enoch Powell euphoniously put it in his 'Water Tower' speech heralding the closure of asylums in the UK, they were the " ... [buildings] which our forefathers built with such immense solidity to express the notions of their day" (Powell, 1961). They embodied the best of intentions yet the impetus for their creation arguably provided the seeds for their downfall. Even in the heyday of their construction, in the late 1800s, the population growth that had in part led to the creation of the county asylum system was leading to overcrowding that became more marked as the 20th century progressed. Similarly, while the ideal of the therapeutic landscape remained intact, buildings that had once been at the forefront of therapeutic design had become outdated and subject to increasing maintenance costs by the 1960s, sometimes over a century after their initial construction.

Against this backdrop and given the intense commitment for widespread closure in Powell's Water Tower Speech, it is perhaps remarkable that any remain as sites of mental health care delivery. Powell, after all, had been direct in his intention: "let me here declare that if we err, it is our duty to err on the side of ruthlessness. For the great majority of these establishments there is no appropriate future use" (Powell, 1961). At the time of his speech there were around 130 asylums in operation in the UK. The majority of these were county (or borough) asylums though some were private or funded through a mixture of public and charitable sources. Closure was anticipated for all but, in the early 1990s, some 90 were still in operation; by the early 2000s the figure had fallen to less than 20.

Identifying current sites of retained mental health care use is far from simple. County or borough asylum status was often obscured by mixtures of funding, the sharing of provision between some counties and boroughs, and the development of new facilities in the first half of the 20th century. Name changes, alterations in jurisdictional responsibility and the exigencies of land ownership and changing responsibilities for the provision for mental health care pose further difficulties. We worked initially with historically-focussed online resources (Table 4.1).

Table 4.1   Online sources about UK County and Borough Asylums

Source
http://studymore.org.uk/4_13_ta.htm
http://thetimechamber.co.uk/beta/sites/asylums/asylum-history/the-asylums-list
http://www.simoncornwell.com/urbex/misc/asylums.htm)
http://www.asylumprojects.org

Note: URLs accurate as of September 2014.

Subsequent searches covered survey papers, books and reports (Taylor, 1991; SAVE, 1995; Lowin et al., 1998, Chaplin and Peters, 2003). Collectively, these sources provided a base list of asylums that have remained open, but were somewhat dated. To generate an up-to-date picture, we reviewed each candidate asylum site on the base list using online search engines, NHS Trust websites and the NHS Choices online hospital search facility. In cases of doubt we followed up with phone calls to local health trusts. It appears that a substantial involvement in mental care delivery persists on some 14 county or borough asylum sites (Figure 4.1).

In settling on this count we made several accommodations. A key word in the final sentence of the previous paragraph is 'substantial'. We excluded cases where only small fractions of former asylum estates are occupied by continuing mental health uses. Such sites include two of the case studies that feature in our discussion of reuse for residential care in Chapter 6: Knowle, the former First Hampshire Asylum, and Graylingwell, the West Sussex County Asylum. We are aware of several others. The designation as a county or borough asylum was also important in our listing. We excluded two of the most significant sites of continuing mental health care use in the UK: the Bethlem Royal and Maudsley Hospitals. These London-based flagship sites for research and care delivery in psychiatry were never county or borough asylums; indeed, the foundation of the Bethlem Royal predated the 19th century county asylum acts by some 600 years. For similar reasons we also excluded the high security hospitals that continue to provide national facilities for offenders with severe mental health problems. These too were outside the county and borough asylum system; they will be given specific attention later in this chapter. Finally, we excluded two further facilities where mental health care currently continues but which were never county asylums under the terms of the

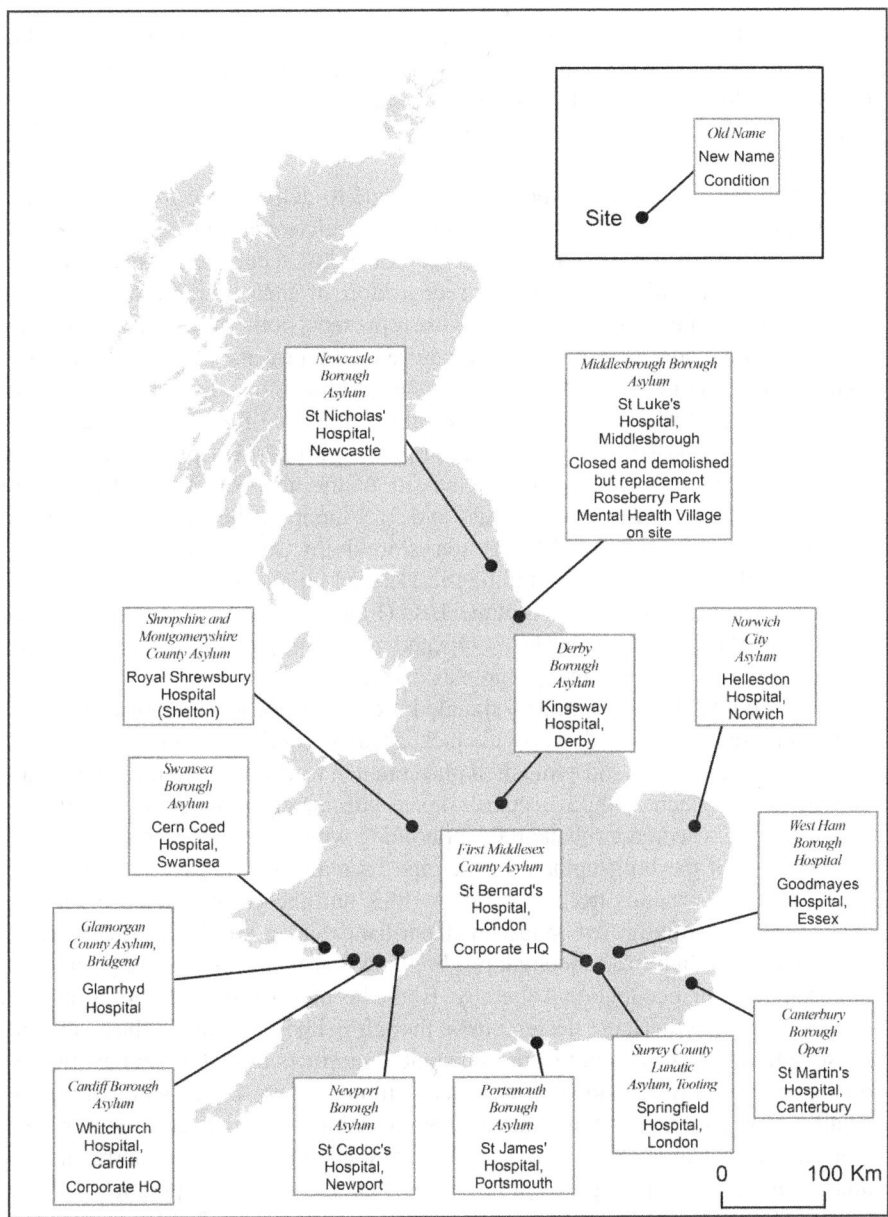

Figure 4.1   Retained County and Borough Asylum Sites in England, 2014

1808 or 1845 Acts. It is an interesting footnote to the saga of retained asylum sites to note that these facilities were ones that underpinned the initial move to the (public) asylum system in the late 18th century. The York Retreat and Bootham Park Hospital are both still involved in delivering care in the second decade of the 21st century.

Not all the identified retained sites continue to deliver care from the original buildings. Several feature NHS administration services as a significant use. Where this has happened, the administrative use is usually accommodated in the more historic asylum buildings, often in recognition of their limited suitability for present-day mental health care. Such re-use represents both a pragmatic use of NHS resources and also a potent visible signature of the power of NHS management. Many of the retained sites also feature new buildings, usually for mental health or related uses but sometimes for other clinical specialities. In some cases this new investment is substantial. For example, St Luke's Hospital (formerly the Middlesbrough Borough Asylum) is the site of the new Roseberry Park facility comprising self-contained ward units and 365 inpatient beds; Shelton Hospital (the former Shropshire and Montgomeryshire Joint County Asylum) features a new 'mental health village' including the 112-bed inpatient Redwoods Centre.

St James' Hospital, Portsmouth, UK (Figure 4.2) serves as an excellent illustration of the points made above. Opened in 1879 as the borough asylum for the City of Portsmouth, it occupies what was once a semi-rural location on the eastern coast of the city. Its development signaled a civic response to population growth and recognition of the need to provide local accommodation for poorer people with mental health problems and avoid their placement in workhouse facilities. Originally occupying a 75 acre (30.35 hectare) site, it subsequently acquired a further 14 acres. This represented a substantial landholding which was, by the end of the 19th century, part of the built-up area of the city. Its main building was supplemented by villas in the grounds from as early as 1883, and the grounds themselves were a characteristic mixture of landscaped parkland and a small farmstead which, alongside workshops for shoemaking, brewing, carpentry and laundry, provided opportunities for occupational therapy. Both the main hospital building and the chapel now enjoy heritage designations, though neither is in the highest category. Preservation orders also protect the substantial numbers of mature and exotic trees on the site. Initially, St James' offered accommodation for 450 patients, of whom a small minority were privately funded. Overcrowding was however evident by 1893 when additional land was acquired. Additional villa developments occurred in 1908 and yet more land was acquired in 1910. By 1926, the resident population had risen to over 1000 and the site included six villas (Thinking Ahead, 2014b).

The advent of community care saw the St James site in essentially the state it had reached by 1930. The resident population had shifted to one in which individuals were predominantly admitted on a voluntary basis, but otherwise the hospital remained largely a self-contained community, albeit one in which there was a certain amount of freedom to enter and depart, and the hospital had something of a reputation for innovations in care. The first major change in the

**Figure 4.2   St James' Hospital Portsmouth, UK, Front
Elevation**

era of community care was the disposal of the hospital farm. The last harvest
was gathered in 1965 and the piggery and dairy operation was closed, with the
farmland being sold to what later became the University of Portsmouth. Today, the
former farm is the location of a student residence complex and the University's
sports ground as well as housing.

The subsequent fortunes of St James' and the trajectory that has led to its current continuing status as a site for the delivery of mental health care can be conceptualised as involving two distinct processes: flexing boundaries and the consequences of governance. The sell-off of the asylum farmlands provided the initial instance of flexing boundaries. Rather than seeing this as the beginning of a process that would culminate in the closure of the asylum, we see it as the genesis of a process of retrenchment in which the mental health function consolidated, concentrated and strengthened at its geographical core, thus facilitating survival. At the same time, flexing boundaries also ensured that the containment of the asylum site was broken, enabling incursions by the surrounding community into former asylum space. This promoted integration with the local community and reduced the mystery of what lay within the walls. Figure 4.3 captures these twin aspects of boundary flexion in action. Therapeutic landscape provides one theme driving these changes. By the 1980s, the cricket pitch, once home to the hospital cricket team, had fallen into disuse. A local club took over the field and it remains their base, no longer a hospital facility but still in its former use. As such it represents both an opening up of the asylum to the external community and the residual presence of a therapeutic landscape in which staff and residents engaged in organised recreation. We can note here, in passing, that this theme of boundary flexing as a facilitator of integration with local communities is also present in case studies presented in this and later chapters: cricket pitches at Knowle (UK) and Porirua (New Zealand), and a rugby field at Kingseat (New Zealand).

A second, and perhaps more dramatic, aspect to boundary flexing was the opening of the asylum boundaries to housing and other institutional uses. Boundaries were first breached during the 1990s with the creation of sheltered housing and, in an echo of themes that will be covered in Chapter 5, educational facilities for children with special needs. The surrounding wall was partially demolished and a new access road was created. In this way the border containment of the hospital was moved to a new inner boundary, separating off new, though related, services from the core hospital and placing them more firmly within the surrounding local community rather than within the (former) walls of the asylum.

Landscape also recurred as a theme in boundary flexing in 1996 when the local community united to preserve the therapeutic landscape of the asylum grounds and secure what had, with the reduced emphasis on containment, become customary rights of access. Though it had formerly been on the edge of the built-up area of the city of Portsmouth, population growth and the expansion of development had ensured that the asylum grounds had become a significant local green resource. A St James Park Trust was formed to secure access to the hospital grounds and, in 2001, an application was submitted to register 1.75 hectares as a 'town green' on the basis that local people had enjoyed access to the area for in excess of 20 years as the boundaries of the asylum had opened up. Portsmouth City Council approved the application but it was subsequently challenged by the Department of Health on the grounds that provisions for town green creation did not apply to NHS land (Hansard, 2002). NHS agreement was eventually gained,

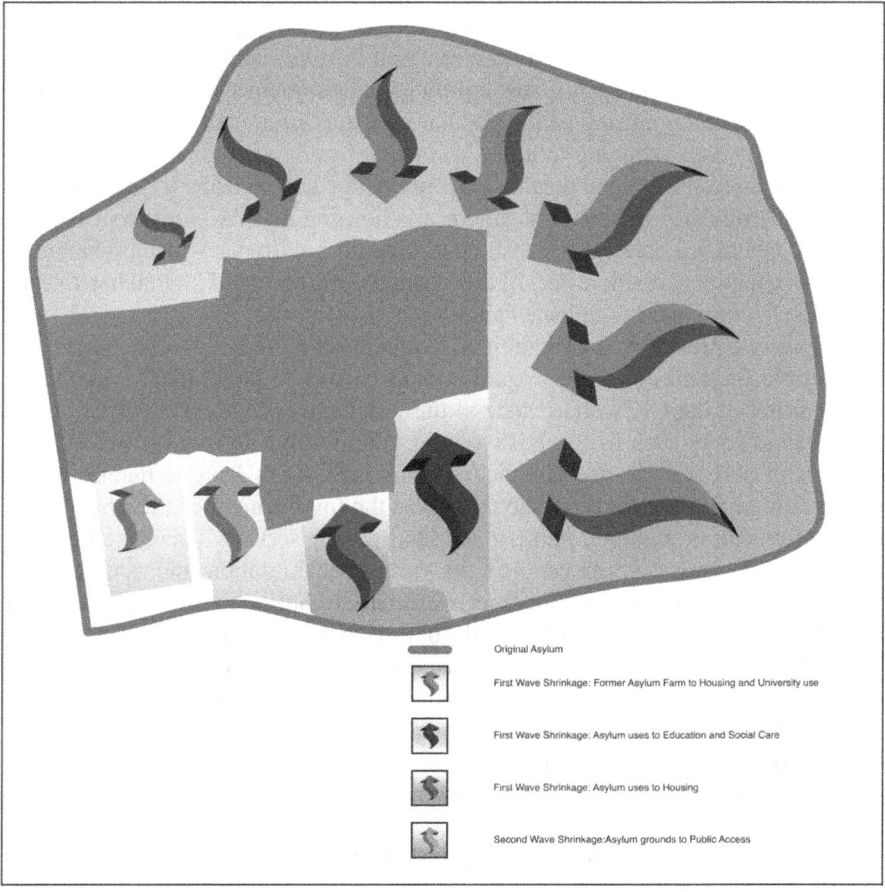

Original Asylum

First Wave Shrinkage: Former Asylum Farm to Housing and University use

First Wave Shrinkage: Asylum uses to Education and Social Care

First Wave Shrinkage: Asylum uses to Housing

Second Wave Shrinkage: Asylum grounds to Public Access

Figure 4.3   Boundary Flexion at St James' Hospital Portsmouth, UK

however, when the council agreed to permit their application for a new mental health development within the hospital grounds. With the Town Green we see the creation of a new border zone of transition in which the established visual appearance of the asylum grounds is maintained but access to them is opened to the community. These 'edgelands' familiarise and expose the local population to a hospital that is increasingly manifest as buildings rather than a mixed environment of building and grounds. At the same time, the disposal of elements of the asylum grounds removes responsibility for maintenance from the hospital authorities.

Housing-related boundary changes continued after the 1990s and indeed have been a constant for the past two decades. The northern boundary of the hospital has moved progressively southward and a housing estate now lies on former asylum land. Further land releases are planned with two former villa sites up for sale in 2014 and building is underway on another site. Current sale brochures describe a former villa site as 'attractively wooded' and note its suitability for 38 houses (Hellier Langston, ND). Local plan provisions suggest a total scope for up to 170 houses on current or future released hospital land though also noting a need for new mental health care buildings (Portsmouth City Council, 2006). Here we see a reconfiguring of the popular construct of mental health care and a symbiosis between changing attitudes to mental health and demand for housing. Integration between mental health care provision and housing is increasingly possible, particularly in situations where it is housing that develops in proximity to mental health provision rather than vice versa, an occurrence which is examined further in Chapter 6. A side-effect of this change is the ability of mental health care provision to survive on established sites.

The second process that we identified in relation to the survival of mental health care on the St James site was organisational change in governance. The broad restructuring of the UK welfare state in the years since 1979 ensured that public sector landowners such as the NHS sought opportunities to involve private and third sector bodies as co-users of facilities and land. At St James a manifestation of this new mixed economy of site users was the involvement of the Shaw Trust in the provision of a garden nursery and landscaping service. This development, 'Greenfingers', was a clear echo of the earlier rehabilitation and occupational therapy roles of the hospital. Workers were largely but not exclusively referred from local mental health services and comprised a plant sales operation and a contract workforce offering a commercial gardening service. Both were specifically set-up to provide training and development to (re)integrate workers into society: "[I]f trainees feel they are ready they can train for NVQ (National Vocational Qualification) level 1 or 2 in horticulture. Greenfingers prides itself on its 'real work' approach. It offers a stepping stone back into work and is self-financing which gives it its commercial edge" (Thinking Ahead, 2014a). Developments like Greenfingers brought people onto the hospital site and took hospital out-patients offsite; it contributed to normalisation and also put in place an organisation with an interest in maintaining its use of the site.

The more specific organisational changes that befell the UK NHS also arguably contributed to the survival of the site. Recent years have seen responsibility for the hospital site and associated service provision fragment between several organisations. Whereas St James' was once the responsibility of the Portsmouth District Hospitals Board, NHS reorganisation after 1979 saw it pass swiftly from being a directly managed unit of Portsmouth and South East Hampshire Health Authority to independent trust status as the main facility of Portsmouth Healthcare Trust, an organisation delivering a range of community-based health services commissioned by Portsmouth and South East Hampshire Health Authority. Neither the Health Authority nor the Healthcare Trust lasted long and at the time of writing St James' provides the headquarters of Portsmouth Clinical Commissioning Group (CCG, essentially the successor body to the Health Authority). At the same time services on site are organised and delivered primarily by the Solent NHS Trust. This fractured pattern of responsibility means that, although a move to community care has remained a long-term and ongoing goal, progress towards hospital closure has been confounded by a diversity of stakeholders. These different stakeholders must now co-ordinate their various needs and ambitions if closure is to happen. Organisations responsible for non-mental health uses (e.g., the cricket club) and contractors (e.g., Greenfingers) are now also involved.

By 2014 it appeared as if such coordination was happening. The Portsmouth City Council local plan and subsequent planned land allocations (Portsmouth City Council, 2006, 2013) set the scene, providing a template for the future of the site in which land to the east of the main hospital building was earmarked for the development of a new health care campus, with the main building released as and when purpose-built accommodation more suited to contemporary mental health care became available. Health care developments (and redevelopments) by the Solent NHS Trust and its predecessor bodies sustaining the mental health use of the asylum site had, by 2014, ensured that the idea of a 'mental health care campus' had become a reality. New buildings catered to brain injury, adult admission and treatment, and the mental health needs of older people, while conversions of former villas addressed needs for detoxification and child and adolescent mental health services. The £8m facility for older people was " ... very well thought through to (sic) helping those with dementia. The place is full of light, which stops them feeling claustrophobic. There are lots of things to do. They're not just locked in a room in front of a telly and forgotten about" (*The News* 1/2/11).

Each of these new or converted facilities, though going by a new name that distances the facility from mental health care, maintains the St James name in its address. Thus the facility for older people is 'The Limes' (recognising nearby trees) but it clearly acknowledges its location on the St James site, consciously embracing stigma. A further new facility was planned to provide a secure place of safety under Section 136 of the Mental Health Act 1983 where psychiatric examination of people detained by the police, either for their own safety or that of others, could take place. The net result of these developments was that St James had, by 2010, assumed a close resemblance to a vision that had presciently been set out in the pages of the in-house newsletter back in the 1960s:

> Let us then look at the possible future of St. James' and mental illness in the year 2000 and beyond. First the buildings themselves; the main building cannot be upgraded indefinitely and must be eventually demolished. Before this comes about it will be surrounded by individual units, each unit with its own specific purpose. These units will be connected by a network of pleasant service roads and the whole set in a series of landscaped gardens. No walls will separate it from the community. Many areas will be set aside for outdoor activities and no doubt a swimming pool will have been built. The units which house the patients, or residents, will be of the villa type and be luxurious by present standards. There will be a gymnasium and a well-equipped centre for social gatherings. There will be classrooms of a sort which will enable the residents to continue with adult education and so broaden their outlook; a centre for occupational pursuits will also be provided. (JBEC, 1965)

There was a clear vision that housing was the appropriate use for the redundant remainder of the site: "The council's local plan has indicated for some time that the site could be developed for a mix of healthcare and housing, so we anticipate the most likely use of any surplus land will be residential" (*The News* 20/2/14). The development of the new mental health campus and the new housing developments on the edges of the site opened up space for debate on remaining land, most notably the main building. A review by Solent Health Care, Community Health Partnerships (the body managing the considerable NHS estate) and NHS Property Services developed proposals for the remainder of the site that were subsequently endorsed by both the Portsmouth CCG and the City Council (NHS Property Services, Solent NHS Trust, & Community Health Partnerships, 2014). The review noted that the majority use of the main building was administrative services, with a key user being the CCG. Less than 5 per cent of the main block was in clinical use and much was empty. With the exception of the new and converted buildings, most of the rest of the buildings in mental health care use on the site were deemed to be no longer fit for purpose: "Maintaining and running surplus and outdated buildings is a significant and unnecessary financial drain on the local health economy" (p. 1). At the same time, less than a kilometre away, £30m had been invested in the CCG's flagship community hospital for Portsmouth, from which a range of other health care services were delivered but where there was also significant vacant space (3000m²) due to the transfer of acute services to another hospital. This nearby but available space was deemed more suitable for contemporary health care provision.

A two-phase plan was proposed (Figure 4.4). First, the Child Guidance Centre, a facility peripheral to the main site and to the main mental health use, would move to the vacant space at the community hospital. Conversion costs at the community hospital would be funded by an interest-free loan from the Department of Health to be paid off by the sale for housing of the Child Guidance Centre land and land associated with three redundant villas closer to the centre of the main St James site. In phase two, clinical and administrative uses in the main building and two connected buildings would move out, to be followed by further housing development. Several of the clinical uses, reflecting the wide community remit of the Solent NHS Trust and the desire to make effective use of the space at St James were, by 2014, not concerned with mental health. These were to move to the community hospital.

The administrative uses were to move to rented office space, which was viewed as cheaper to maintain and service than the current St James-based provision. A mental health day centre was gazetted for closure in 2014 with re-provisioning through a contract with an independent off-site day centre (Portsmouth City Council, 2014). The remaining mental health uses were scheduled to move to a new building on the mental health campus by 2016/17, ensuring continued provisioning on the St James site into the future but with the main building converted to apartments and the remaining land developed for approximately 250 new houses (NHS Property Services et al., 2014). A subsequent proposal saw the

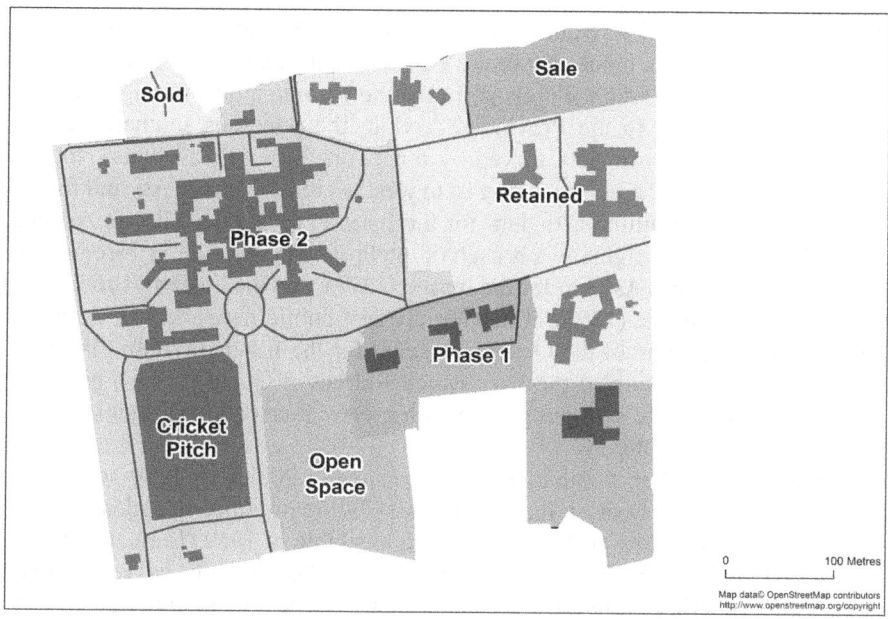

Figure 4.4   Future Plans for St James' Hospital Portsmouth, UK

sale of adjacent University land (formerly the asylum farm) for housing emerge as a possibility.

St James' provides an indicative case study of how mental health uses have been retained on a former asylum site in the UK. The retention process is not only one of inertia but also boundary adjustment. Both of these spatial processes have involved services concentrating into specific parts of the former site. However, there is also an organisational dimension, with the co-location of other service providers enabling continuing partial use of buildings and sites for mental health uses. The organisational complexity arising from this acquisition of additional functions, each with their governance structures and reporting lines, ensures that proposals for change require complex consultation processes. Moreover, the broader move to community care signified by the opening up of sites to external users undoubtedly facilitates acceptance of continuing mental health uses. While such uses may continue, it is also clear that, as a consequence of the diverse processes that enable survival, the retention of mental health care across the whole of a former asylum site is unlikely. We will return to this idea of partial retention as normative in the concluding section of the chapter.

Retention and the Provincial Asylum System in Ontario

The provincial asylum system of Ontario had its origins in the same Victorian-era debates that spawned the English county and borough asylum system. The (at the time) single Provincial Asylum for Ontario, sited in Toronto, was completed in 1851 and constituted a model for future development. Indeed, the second half of the 19th century witnessed the propagation of the Toronto model through Ontario, such that in addition to the Provincial Asylum the populous southern portion of the province was, by 1900, served by a network of public asylums. Notable amongst these, and progressing from east to west, were the London Asylum for the Insane (1870), the Hamilton Asylum for the Insane (1876), the Mimico Asylum for the Insane (1890; originally a branch of the Provincial Asylum in Toronto),the Rockwood (Kingston) Asylum for the Insane (1877), and the Eastern (Brockville) Asylum for the Insane (1894). The 19th century public network was extended through the completion of the Ontario Hospital for the Insane (Whitby) in 1919, while the earlier opening of the Homewood Retreat in Guelph (1883) provided an option in the private sector for those with greater financial means (Joseph and Moon, 2002; see Chapter 3).

The Ontario asylum population peaked at nearly 20,000 in 1960 and began to decline thereafter (Dear et al., 1979). By 1979 the province had closed 7,000 of the 11,000 psychiatric beds that had been available in 1959 (Lurie, 2005). As in England and Wales, these data should not be taken literally as a barometer of deinstitutionalisation. In the 1970s and 1980s the decline in capacity in what had generally become known as psychiatric hospitals was accompanied through transinstitutionalisation in the form of the transfer of responsibilities for in-patient care from psychiatric hospitals to newly-established psychiatric wards in general hospitals (Sealy and Whitehead, 2004). It was not until the 1990s that community care clearly emerged as the dominant modality in the provision of mental health care services in Ontario (Hartford et al., 2003).

We noted in Chapter 3 that the private Homewood Retreat, now known as the Homewood Health Centre, continues to function as a psychiatric hospital. By taking provincially-funded patients, it became a *de facto* part of the public system in the late 1960s (Joseph and Moon, 2002). It is thus, in its own specific way, a survivor of the 19th century asylum system. What, though, of the six definitively public psychiatric asylums mentioned above? In all cases other than the Mimico Asylum for the Insane (latterly the Lakeshore Psychiatric Hospital) mental health care continues to be delivered at each of the listed southern Ontario locations. Site visits in 2013 revealed that in-patient care is still among the suite of services offered. Crucially, however, this provision is not in the original asylum buildings. We discern two phases in this redevelopment of sites. Initial redevelopment in the 1950s and 1960s saw the transfer of some in-patient services, often incorporating forensic units, from the original (often Victorian-era) buildings to relatively modest but custom-built buildings. Arguably, this on-site transfer from the hitherto main asylum buildings represented a literal and symbolic retreat from the confidence expressed through the imposing monolithic architecture of the Victorian asylum. Nevertheless, it represented a form of retention, at least in terms of site.

A second phase of retention in Ontario is ongoing, and features the extensive actual or planned re-development of several sites. New and imposing buildings are proclaiming a new confidence in the efficacy of hospital-centred mental health care networks on a scale significantly larger than the mental health campuses that had been developed in the 1950s and 1960s. We see this locational continuity as indicative more of the difficulty of finding new locations accepting of mental health services and less of the enduring suitability of initial asylum sites. It also represents a fiscally conservative approach to maintaining the land base of the system. However, as we discuss below, this does not mean that portions of the (usually extensive) estates of the asylum could not be disposed. Moreover, in Chapter 5 we present a case (that of Lakeshore in Toronto) in which the Province disposed of a site, albeit after considerable delay.

Ironically, it was the 'newest' of the original psychiatric asylums, the Ontario Hospital for the Insane in Whitby, that was first to be re-built. Since its opening in 1919, the Whitby facility has gone through four name changes, the first three of which paralleled those across the province's health care system. In the early 1930s, it was re-named the Ontario Hospital, Whitby, and in the late 1960s, it became the Whitby Psychiatric Hospital. Between 1994 and 1996, the facility was totally re-built on part of the original site and was officially re-opened as the Whitby Mental Health Centre in 1996. In 2009, the Centre was re-named the Ontario Shores Centre for Mental Health Sciences (Ontario Shores, 2011). Ontario Shores is a psychiatric hospital in all but name; it has 330 beds, incorporating two forensic units on site, and supports a network of hospital- and community-based services throughout its region. Recent developments of housing on the periphery of the hospital, on what was formerly part of the asylum estate, have brought hospital and community into close proximity. Only parking lots now exist as a buffer between the hospital and the neighbouring housing development. During our site visit, a hospital respondent reported fears that the province might chip away further at the hospital's estate in an echo of the boundary flexing process identified in the St James case described earlier.

Elsewhere, ambitious plans are in place for the on-going re-development of the Centre for Addiction and Mental Health on the site of the former Provincial Asylum in Toronto and new facilities are near completion in London and Hamilton, the latter on the site of the former Hamilton Asylum. Each has a strong relationship with medical teaching programmes at local universities. A substantive re-build has been approved for the Kingston site, but construction has not yet begun. Services continue at the former Eastern Hospital for the Insane in Brockville, now renamed as the Brockville Mental Health Centre with facilities for 200 inpatients.

While these new and continuing facilities are reminiscent of the asylum in their scale and, in all cases but London, located on the sites of former psychiatric asylums, they do not automatically trigger remembrance of the asylum as a modality of care. In no small part, this is a result of contemporary naming conventions in which even reference to terms such as 'psychiatric hospital' is assiduously avoided. For example, in Hamilton, we are told that the "St Joseph's Healthcare Hamilton West 5th campus is home to the regional specialized Mental Health Services for Central South" although later in the same description it is referred to as a facility with a capacity of 305 beds (St Joseph's Healthcare Hamilton, 2014). Visitors are struck first by the considerable size and obvious sophistication of the new hospital, and second by the more modest (1950s and 1960s) buildings associated with the Centre for Mountain Health Services that assumed the responsibilities of the former Hamilton Psychiatric Hospital in 2000 (Figure 4.5). Only after that would a visitor be further struck, or perhaps perplexed, by a few of the original asylum's secondary buildings that survive on the site. Indeed, of the Ontario sites, it is only at Kingston that the large and imposing core buildings of the former asylum are still intact. They sit, boarded up and fenced in, between the shore of Lake Ontario and the current (and relatively modest) mental health centre. Their fate, as the best surviving example of Victorian asylum architecture in Ontario is less than certain.

**Figure 4.5   St Joseph's Hospital West 5th Campus,
                 Hamilton, Ontario Canada**

The Ontario case has certain parallels with the English (and Welsh) county and borough asylums. Renaming is evident, as is partial reuse of former sites. What is largely different is the continuity with the former Provincial Asylum system. Survival in some form, in contrast to England and Wales, is the rule rather than the exception. There are analogous examples of the development of new facilities in the form of mental health campuses on former asylum sites in England. However, the commitment in Ontario to inpatient residential care facilities of significant size appears to signal a greater role for separate hospital facilities for people with mental health problems. These facilities, we stress, do not represent a return to the asylum; rather they should be understood as a counterpoint to earlier moves to integrate inpatient mental health care into general hospitals and a recognition of the scale of demand for inpatient care. In so far as there are continuities with the asylum, they lie in the continued separation of care for people with mental health problems from the general hospital system and a co-existence of institution and surrounding community.

Exceptional Situations

Having examined the themes underpinning the survival of mental health care provision on former asylum sites in England and Wales and in Ontario, we turn now to two exceptional situations: forensic mental health care and the continued use of asylum sites as museums commemorating their former use. Forensic care is a continuity of use based on the needs of a particular client group and the societal response to that population. Museum uses create fossilised palimpsests of the former use and ensure a continuing – if static – onsite presence of the asylum care modality. Both are replete with prospects for the remembrance of the asylum.

Forensic Mental Health Care

As we noted earlier in this chapter, the move to community care always carried a recognition that there were groups of people who would have a continuing need for inpatient mental health care. One such group were individuals whose mental health status was associated with their conviction for a criminal offence. In-patient detention in secure residential 'forensic psychiatric' facilities continues to be the dominant treatment modality for this client group.

In Ontario, the Penetanguishene Asylum for the Insane, is now the Waypoint Centre for Mental Health, serving its local Georgian Bay area but also providing maximum security forensic mental health services to the whole of Ontario. In the UK, a system of low, medium and high secure facilities provide for individuals who were historically called the criminally insane and latterly labelled as mentally-disordered offenders; the majority of patients are detained under the Mental Health Act, 1983 (and the equivalent Act in Scotland) for their own safety and the safety of the public. At the apex of the UK system are the four high secure hospitals: Broadmoor (opened 1864; Figure 4.6), Rampton (1912), Ashworth (c.1970 as an amalgamation of two predecessor facilities) and Carstairs (1957, serving Scotland and Northern Ireland). These are highly specialised facilities with a singular focus on forensic psychiatry. Medium and low secure services are often provided on sites where there are also other non-forensic services but can be stand-alone, particularly where they are provided through contracts with the private sector. There are several instances where medium or low secure facilities constitute a residual presence on otherwise closed former asylum sites; our case study of Knowle Hospital in Chapter 6 provides an example where a medium secure facility continues on the edge of a site otherwise redeveloped for housing.

The scale of the forensic mental health care system is not insubstantial. The latest figures for England suggest that there some 680 high secure beds, 2800 medium secure, and 2500 low secure (NHS Commissioning Board, 2013). At the low secure end, most provision is in units of less than 30 beds. The three high secure hospitals are considerably larger, with Broadmoor currently housing approximately 250 people. This scale of operation raises several issues. As Mullen (2000, pp. 308–9) has argued: " ... much of the progress in the organisation and delivery of general mental health services has passed forensic services by. The anachronistic and unforgivable giant high-security hospitals still dominate not just British forensic mental health services but those of much of the Western World. Community-based and rehabilitative services are often rudimentary or non-existent". This quote highlights a rather obvious contrast between the abandonment of the asylum in general psychiatry and its retention in forensic psychiatry (see also Bluglass, 1992). This is perhaps to be understood in the light of public concerns over dangerousness and the involvement of the criminal justice system but it also points to an ongoing conflict between the therapeutic and the containment expectations associated with forensic facilities. There are historic echoes of pre-asylum health care, at which time there was sometimes little differentiation between 'mad' and 'bad'. There are also reminders of the substantial prevalence of mental health problems in present-day 'mainstream' prisons, exacerbated by poorly coordinated community care that often finds it difficult to place people with more complex mental health problems (Trieman and Leff, 1998).

If the nature of the client group provides the central explanation for the survival of asylum-type facilities in forensic psychiatry, it also underpins the visual presentation of the facilities, and in this we see implications for the remembrance of asylums themselves. The conflict between containment and care that impacts on the therapeutic regime is evident in the spatial signature of the facilities. Internally there might be commitment to care and rehabilitation but externally the accent is on confinement, high fences and prison-level security standards. From the outside, the facilities appear as prisons rather than hospitals; they feature levels of containment and security far in excess of those ever used in asylums. Nonetheless, they are mental health hospitals and have been since 1948 when they were transferred from the Home Office to the NHS. In their governance, however, lies a secondary reason for their continued existence. Following their transfer to the NHS, the three English Special Hospitals retained separate status outside the mainstream NHS system along with distinct management arrangements. Only since 2001 have they been managed by local NHS Trusts with responsibility for other (non-forensic) mental health services. While they undoubtedly serve a very distinct and specific need, it may therefore also be that their survival as large-scale residential facilities owes something to their lengthy period of governance outside the structures that brought change to the majority of mental health services in preceding decades.

Figure 4.6   Broadmoor Hospital, UK

The interplay of correctional, justice and health services evident in the UK case is also present in the far more recent development of forensic provision in New Zealand. As we noted above for the UK, mental health problems are extremely prevalent in prisons. In the late 1980s, the New Zealand prison population experienced systematic under-provision of forensic care, resulting in high rates of suicide (Skegg and Cox, 1991). This unfortunate trend prompted the government to initiate the Mason Inquiry in 1988, although the direct impetus for the investigation was an incident in which a mentally ill ex-offender stabbed a fellow resident at his hostel and a person in the street. To conduct his inquiry, Judge Mason and fellow committee members travelled both within New Zealand and internationally (Capital and Coast District Health Board, 2007). The key outcome was the establishment of six (now five) Regional Forensic Mental Health Services to complement the single National Secure Centre at the Lake Alice Psychiatric Hospital, itself only dating from 1965 (Mason et al., 1988; Chaplow et al., 1993; Radio New Zealand, 2006). In essence, the separate and distinctive forensic hospital network that had evolved historically in the UK thus emerged over a short time period in New Zealand largely out of the deficiencies of a prison-based forensic system.

The eponymous Mason Clinic (Figure 4.7) was established to provide forensic psychiatric care to offenders experiencing mental illness from across the Auckland and Northland regions (Waitemata District Health Board, nd). It was, and at present continues to be, located on the grounds of the former Carrington Hospital and on the perimeter of the successor use, Unitec, in Mt Albert, Auckland (see Chapter 5). Like Knowle, near Southampton in the UK (see Chapter 6), it thus exemplifies the situation where forensic care provides a continued mental health care use on a former asylum site. Following the announcement of the imminent closure of the National Secure Centre at Lake Alice in 1998/9, the Mason Clinic is now the largest inpatient forensic psychiatric facility in New Zealand and provides both inpatient beds and community-based services. It has experienced consistent growth and is often called upon to provide care to people with severe mental health problems transferred from prisons that do not have adequate mental health facilities, as well as from other regions. In January 2003, it was described as at being in crisis when the Department of Corrections planned to place 12 additional Auckland prisoners in the already-full clinic (Waitemata District Health Board, nd).

Figure 4.7   The Mason Clinic, Auckland, New Zealand

In an echo of earlier practice of co-location in psychiatric asylums such as Tokanui (Joseph and Kearns, 1996), in June 2006, the Waitemata District Health Board opened a secure intellectual disability unit at the Mason Clinic known as the Pohutakawa Unit. This purpose-built 12-bed facility was established in response to a gap in service provision for offenders experiencing intellectual disability. The need for this innovation was identified by the Mason Clinic's then Clinical Director (Radio New Zealand, 2006) who explained that the unit aimed to develop

" ... a culture of building bridges, bearing in mind community safety". A 10-bed Māori forensic unit, Tane Whakapiripiri, was also opened at the Mason Clinic in 2006, aiming to provide a *kaupapa Māori* (Māori philosophy) approach to mental health, incorporating Western psychiatry and aspects of Māori-specific intervention (e.g., education regarding ancestry and spirituality alongside pharmacological interventions) (New Zealand Doctor, 2006). A further six on-site inpatient units complete the current provision of services at what is described as an "integrated forensic mental health service" for mentally-ill offenders. (Waitemata District Health Board, nd; Tepou, nd).

In 2012, the *New Zealand Herald* (12/11/2012) reported that 104 patients were receiving care at the Mason Clinic. The then clinical director explained that "For those requiring lower levels of security, gradual access to the community is facilitated, at first this being escorted by staff". Perhaps remarkably, the first few hundred metres of such a re-entry to 'the community' necessarily takes the reintegrating forensic patient through the undergraduate campus of Unitec (see Chapter 5). While the (continuing) presence of mental health care on the former Carrington Asylum site may be explained by the availability of land, a legacy of a secure ward at Carrington, and the desire to maximise the utility of the asylum land holding, it is increasingly incongruous in the corner of a higher education campus. It was perhaps unsurprising when, in 2014, the management of Unitec suggested that the clinic might like to vacate the site to facilitate a plan to sell off surplus UNITEC land for housing (*New Zealand Herald*, 18/6/2014), implying a belief that the stigma associated with the Mason Clinic might be a barrier to the successful re-development of the site.

The Asylum as Museum

The psychiatric museum is arguably the ultimate artefact and materialised memory of the practice of mental health care on asylum sites. As an instance of continuing mental health care use on former asylum sites, museums fall clearly into our 'exceptional' category. While they are not a form of service provision, they commemorate service practices and record the asylum past, presenting it to an audience for whom it is likely to be increasingly unfamiliar. In this sub-section we examine briefly the development of hospital museums generally before profiling two key examples: the Porirua (New Zealand) and Glenside (UK) Hospital Museums. We stress how, on asylum sites that are now wholly or partially closed, the process of remembering the asylum past is facilitated though memorialisation that is materially and symbolically embodied by the museum. Hence museums become both a house of memories and a place from which to de-code the present built environment of mental health care.

Despite considerable concern for its own history (Johnston and Sidaway, 1997) and even the preservation of its archive (Johnston and Withers, 2008; Withers, 2006), the academic discipline of geography has had little to say about museums. A partial exception is an agenda setting paper by Geoghegan (2010). Additionally, useful analogies can be drawn from geographical research on commemoration and remembrance at stigmatised sites, notably concentration camps (Azaryahu, 2003; Charlesworth, 1992; Giaccaria and Minca, 2011; Legg, 2007). Closer to our subject matter, Cooke and Jenkins (2001) offer reflections on the transformation of the former Bedlam asylum London into its current role as the Imperial War Museum. There is a similar situation of relative neglect but useful parallels in other disciplines, with the clear exception of heritage studies where notable contributions have shown how museums have attempted to problematise the past, recover alternative voices and also preserve vanishing artefacts in ways that vary from focused facilities to individual exhibitions (Arnold, 1996; Candlin, 2012; Sandell, 2003; Stearn et al., 2014).

There has been little research on museums devoted to health care or hospitals though there is a long history to such facilities. Recent studies have tended towards critical museology, and for our purposes offer useful insights. Sandell et al.(2013), for example, offer reflections on the presentation of disability in museums and their collection includes coverage of mental health care. Others have considered how museums can facilitate recovery through enabling people with mental health problems to handle and respond to artefacts (Dodd, 2002). Most usefully for us Coleborne and MacKinnon (2003, 2011) have been concerned directly with asylum museums and their development, arguing that: " ... psychiatric practices as well as those who have been made subject to its regimes become more visible when we consider both the material and visual cultures produced through and by psychiatric institutions and also the later representations of these as they are embodied in museums, collections and displays" (2011, p. 4). In other words, the asylum museum can provide a lens for continued understanding of past mental health care practices and, in that sense, the use of former asylum sites for museum purposes forms a distinct but unusual continuing mental health use.

Drawing from the above and our own observations, it is possible to identify two themes historically common to the hospital museum. First, there is a strong emphasis on the preservation of artefacts that symbolise past practice. Thus collections feature surgical equipment, uniforms and portrayals of key staff. By current museology standards these are rather traditional concerns, though there are important exceptions. Facilities such as the Thackray Medical Museum in Leeds (http://www.thackraymedicalmuseum.co.uk), coincidentally housed in a former workhouse, are fully-tuned to contemporary museum norms and incorporate activities, costumes, smellscapes and educational programmes. Second, hospital museums frequently owe their origins and their survival to the efforts of individual enthusiasts, often former members of staff. Interests fostered during a working life spill over into retirement activity, preserving familiar reminders of a professional past and documenting the former working environment. A contemporary extension of this aspect of the hospital museum is the creation of websites where staff members share memories of their former institution.

The two themes of the artefact and the enthusiast are evident in the case of the asylum museums on which we now focus. Both the Porirua (Wellington,

New Zealand) and the Glenside (Bristol, UK) museums are relict presences on the sites of former asylums. As mentioned earlier, the Porirua site contains continuing mental health care uses, though services are offered from new buildings on the site of the former asylum. Glenside now operates largely as a campus of the University of the West of England although the site also contains a regional secure unit. By the 1970s, most residents of the Porirua asylum had been placed in community-based care and the last functioning ward – F Ward – was closed in 1977 as it had become both uneconomical and unfit for use (Porirua Hospital Museum, nd). The Porirua Hospital Museum was opened in 1987 in the former F Ward (Maclean, 2009). Open only for one afternoon each week, its collection was (and is) typical: "Scary medical devices sit in glass cabinets, while a skeleton leers from a cupboard. There is a seclusion room with a straitjacket hanging at the ready and a dentist's surgery where rows of teeth were routinely taken out" (*Dominion Post*, 22/10/13).

Figure 4.8   Porirua Museum, Wellington, New Zealand

Coleborne (2003) examined the historical meanings of such artifacts, asking why 'relics' of an often unpleasant past were preserved. She identified their importance as a record of the asylum experience and the value of interpreting history 'inside the space of the museum'. While acknowledging the basis of such interpretation, it also seems to us that such an emphasis on relics, as opposed to a focus on systems of care and on caring itself, plays into a popular culture view of the asylum. We will return to this caveat in the concluding section of the chapter.

The museum continues to operate today in the remains of what was the original asylum. A sun shelter, originally constructed to protect asylum patients at risk of damage to their skin due to medications, still stands on the museum grounds (Maclean, 2009). The museum collection is a record of the development of mental health services in New Zealand over the past 150 years and is the only such collection in the country. It continues to rely on a part-time curator and volunteers, as well as the Incorporated Society, Friends of Porirua Hospital Museum, which was formed in 1997. A Trust with representation from the Friends of Porirua Hospital Museum, Porirua City Council and the Capital and Coast District Health Board, was formed in 2006 to oversee the museum. The Trust has a 10 year lease on the museum building and grounds.

The future of the museum is far from secure. In 2008, the Porirua City Council granted $15,000 to the museum to fund replacement of the F Ward roof and assist the repair of water damage to the building, and the museum also received funding from the Lotteries Commision and undertook fundraising efforts. However, by 2012 it was reported that the museum building was on the market and could soon be cleared away, allowing the hospital land to be prepared for possible settlement of a Ngati Toa Waitangi Treaty claim (*Dominion Post* 10/7/12). The report suggested that the building was in a ruinous state and represented a liability for the Crown. The following year the *Dominion Post* (22/10/13) reported that, due to its leading roof, mould-ridden rooms and its generally decrepit state, the former mental asylum building faced threat of closure. On account of its increasingly poor state, the Capital and Coast District Health Board was reviewing the future of the museum which, though officially a historic building, was experiencing a lack of financial support that meant the museum's Board of Trustees were struggling to cope with the upkeep of its facilities.

Glenside Hospital, the former Bristol Lunatic Asylum, closed in 1994. Hallows (2011, p. 10) notes a classic genesis: the museum " ... only exists because a consultant psychiatrist, Donal Early, started squirrelling away items that his employers no longer wanted – old ECT machines, syringes, the odd straightjacket – on the balcony of the dining hall". It is perhaps indicative that Early later went on to write the definitive history of Glenside (Early, 2003). On the closure of the asylum, the museum moved to the Grade II listed former chapel (Figure 4.9). Beyond its collection of artefacts and its not inconsiderable web presence (www. glensidemuseum.org.uk), the museum features exhibitions on the artist Stanley Spencer, who was an orderly at the asylum, and paintings by a former patient recording the everyday life of the asylum in its final decades (Ramsey et al., 2008). The museum also incorporates material on other hospitals, enabling coverage of analogous facilities from the local area specialising in learning difficulties and brain injury. Somewhat less relevant, and indeed incongruous is the presence of a fossil collection and a horse's tooth, further indicators of the serendipitous approach to assembling the museum's collections.

Figure 4.9   Glenside Museum, Bristol, UK

Like Porirua, Glenside relies on a corps of volunteers to ensure its future. Unlike Porirua it has, however, recently been refurbished. The initial move to the former hospital chapel required considerable effort from the museum volunteers as the building was in a bad state of repair. Though this was largely successful, the museum had to be closed in 2009 to enable electrical and roofing work to be carried out. The University of the West of England invested £40,000 as the museum's landlord (*Bristol Post* 21/9/10). Since reopening, the museum has embraced new ambitions. As the museum director indicated: "Originally when we came here, we literally exhibited items in rows on the church pews … But now we have been able to install proper exhibition spaces, which we hope will allow us to educate more people about the history of psychiatric care on the site". Progress to this end has been enhanced by two large grants from the Heritage Lottery Fund to collect memories of Glenside Hospital from former patients and staff with the aim of enhancing understanding of the history of mental health care. Additional funding has been secured from the Esmee Fairbairn Collections Fund to explore the First World War history of the hospital, at which time it was requisitioned as a facility for the treatment of wounded soldiers.

Neither Porirua nor Glenside is unique, though within the New Zealand context Porirua could claim that distinction. In the UK, among others, the former West Riding (Yorkshire) Asylum is memorialised in a digital archive (http://www.wakefieldasylum.co.uk) as well as a physical museum – The Mental Health Museum   http://www.southwestyorkshire.nhs.uk/quality-innovation/mental-health-museum). The latter aims to " … reduce prejudice towards mental health problems and be candid about where the evolution of mental health services have progressed and where it did sometimes fail". As at Glenside, the beginnings of the museum can be traced to a single individual, a former Secretary to the asylum who took a keen interest in recording and researching its history (Ashworth, 1975). The Mental Health Museum differs however in that it is no longer based on the premises of a former asylum and, as such does not contribute to our theme of site survival.

Elsewhere, there are psychiatric museums at, among other sites, the Oregon State Hospital (http://oshmuseum.org), featuring a 'One Flew over the Cuckoo's Nest' exhibition commemorating the filming of the movie at that institution. In Canada, at the former Ontario Shores Whitby Provincial Hospital, Ontario, there is a display that celebrates a tradition of caring at the Whitby site and traces the evolution of treatment from the asylum period to the mid-1990s. Partly as a consequence, the exhibit is seen by some staff and patients as more of a reminder of the stigma associated with mental illness than as a signifier of the progress made in its treatment. Beyond the English-speaking world, other museums have been documented in the Netherlands (Löwenberg-Doornbos and Freedman, 2008) and more widely across Switzerland and Germany.

Conclusions

This chapter has drawn on archival and documentary research supplemented by field visits to asylum sites where there are significant continuing health care uses. The majority of these uses are for the provision of psychiatric care, although we have identified instances where there has been such survival alongside diversification into other health care uses or into health care administration. We have noted a range of issues that are implicated with the retention of asylum sites – renaming, complexities in health services management and a tendency towards a spatial focusing onto smaller fragments of former asylum estates. The latter trend raises a question that we encountered early in our research on retention: how partial can retention be before it no longer constitutes continuity and a related ability to convey remembrance of the former asylum? We have also noted how retention is often coupled with a repackaging of provision into newer buildings, some of which can be of considerable size. There are additionally, of course, implications for the remembrance of the asylum. Renaming may obscure the purpose of sites and buildings for some, but in proximate communities it is likely that the nature of 'their hospital' will be known and remembered. Renaming has potentially greater impact when daily activity has disappeared. In their absence, evidence of the asylum past can be obscured through the processes of strategic forgetting and selective remembrance introduced in Chapter 2. We will show this to be the case in re-use of sites for tertiary education (Chapter 5) and especially housing (Chapter 6).

The flexing of boundaries, which we observed at several sites and described in detail for St. James' Hospital (Portsmouth), results in housing and other developments encroaching progressively on residual core sites of service delivery.

Notwithstanding the possibility of such co-location of housing and delivery of mental health care services becoming an issue in the future, surrounding communities now seem currently to be more concerned with the progressive loss of green space from asylum sites under further development than with the retention of services. These green spaces have acted historically both as a therapeutic environment as well as a means of guaranteeing 'asylum' from a troubling world, but have increasingly been seen by communities as a common resource. This emergent importance of asylums sites as resources for local communities, will be re-visited in later chapters, but suggest that, in some circumstances, the impacts of stigma can be mollified if not overcome by community attachment.

We see very different, and stigma-laden, implications for remembrance flowing from the exceptional cases we considered – the retention of some sites for forensic mental health care and the transformation of remnants of others into museums memorialising the asylum past. We see both as promoting a very uneven remembrance of the psychiatric asylum. Contemporary forensic facilities, especially those in older buildings and suffering as many do from over-population, create a recollection of the asylum as a highly custodial facility. In truth, such memories are less reminiscent of the asylums than they are of the prisons from which they saved the mentally ill in the 19th century. Similarly, with important exceptions, museums convey an image of the asylum that is less about caring than about odd and, often to contemporary eyes, cruel ways of treating mental illness. Displays of electro-shock machines and lobotomy tools all-too-often trump images of caring staff and untiring attempts to rehabilitate. Such correlation with popular culture imagery of the psychiatric asylum as a 'place of horror' will be considered further in Chapter 7 in connection with abandoned asylums and their exploitation.

Chapter 5

From Asylum to College Campus: Memory and Remembrance in Transinstitutionalisation

In this chapter we examine the fate of transinstitutionalisation, whereby former asylum sites and buildings are re-purposed as new institutional uses that meet contemporary societal needs and priorities. To exemplify this re-purposing within the public sector, we examine the transformation of former psychiatric asylums into educational facilities. Our focus on repurposing for education flows from the importance of policy considerations first noted in Chapter 2. In settling on our focus within the broader fate of transinstitutionalisation we are responding to the coincidence of asylum closure with the aggressive expansion of tertiary education provisioning across virtually all developed economies. We are particularly interested in the ways such conversions have proceeded in the face of the long shadow of stigma associated with asylums and in the measures undertaken by new owners to purify space and mask the former use – which we refer to as strategic forgetting. We are also interested in identifying examples of deliberate memorialisation which stand in contrast to this dominant trope of distancing from the past.

Our principal case studies in this chapter are the transformations of Lakeshore Hospital (Ontario, Canada) into the Lakeshore Campus of Humber College in Toronto and of Carrington/Oakley Hospitals (New Zealand) to the Mt Albert Campus of 'Unitec', a higher education provider in Auckland. Supplementary insights, from the former Challinor Hospital (Australia) (currently the Ipswich Campus of the University of Queensland) and portions of the former Kalamazoo State Hospital (USA) (now Western Michigan University), are used to reflect on the generality of key results. Collectively these examples allow us to argue that the 'successor' contemporary educational use of asylum sites, with its implicit orientation towards learning and catering for young people, generally leads to suppression, where possible, of evidence of the former asylum. This de-emphasising of past use becomes problematic, however, when buildings are retained by choice or necessity and hence memorialisation becomes a tacit, if unintended, presence through the built form of the new tertiary educational campus. In both our primary examples of metamorphosis into higher (tertiary) education institutions we note a broad support for the new use from surrounding communities. Local residents see a guaranteed preservation of green space and employment that would not occur to the same extent if the former asylum was re-purposed as housing.

We pose two questions: what conditions facilitate a transition to educational re-use, and how is former asylum use remembered and memorialised in the successor (educational) context? Through the chapter, we explore the variable responses to the shadows cast by stigma and the vilification of asylum. As introduced in Chapter 2, we distinguish between memorialisation (material reminders on site) and remembrance (narratives of past use). Former asylum sites, we contend, are attractive for educational users because of their campus-like settings (trees, lawns, gardens, driveways) dispersed sets of often 'character' buildings, and (now) suburban locations. For city residents and planners these site are attractive, we argue, because replacing one institutional use with another keeps the site green, brings employment, and retains semi-public access. We contend that this acceptability of educational re-use is enhanced by the fact that memorialisation is often strategically low-key and remembrance more personal and individual. The net result, we propose, is a relict landscape that speaks to the transcendence of stigma despite the relatively recent demise of the asylum (Kearns et al., 2012).

The re-use of asylum sites and buildings is a challenging process, raising questions such as: how is the past to be transcended and how do new uses confront or assimilate the stigma of the former use? As noted in Chapter 2, in addressing these questions we draw insights from work on the significance of symbol and memory within the landscape of health care (Kearns and Barnett, 1999; Gesler and Kearns, 2002; Moon and Brown, 2001) and from studies of the afterlife of stigmatised sites (Charlesworth, 1994; Van Hoven, 2006). In so doing, we seek to see 'beyond the scene' (Stephenson et al., 2010) that confronts the contemporary user or visitor; to understand the processes involved in honouring history and maintaining heritage while minimising the current impact of a stigmatised past use.

The remainder of the chapter is organised in four sections. The first briefly considers memory and remembrance in the built environment, elaborating on the theoretical background introduced in Chapter 2. Second, we present the Canadian and New Zealand case studies, considering first the journey toward transinstitutionalisation and then evidence for the memorialisation and remembrance of the previous, asylum use. Third, drawing on the Australian and United States examples as well as the two primary case studies, we reflect on the general opportunities and constraints associated with the educational re-use of psychiatric asylums. We conclude by considering the extent to which memorialisation and remembrance are subsumed within an imperative to re-brand sites and buildings in a strategic forgetting of former stigma.

Memory and Remembrance in the Built Environment

A burgeoning literature in social and cultural geography addresses the question of how collective memory and remembrance are produced by, and reflected in, the built environment (e.g., Hoelscher and Alderman, 2004; Till, 1999; Johnson,

2008; Gould and Silverman, 2013). Indeed, as we discussed in Chapter 2, the links between landscape and memory are pervasive (Schama, 1995).

Tensions, however, commonly exist between what is remembered and memorialised and what is not. The dimension of political contestation has received particular attention. Waitt and McGuirk (1997), for instance, demonstrated how Miller's Point on Sydney's waterfront was selectively incorporated into the city's cultural tourism landscape through constructions of national identity that suppressed, trivialised or silenced other readings of the site. At a national scale, Forest et al. (2004) argue that even when countries have experienced dramatic reductions in centralised state power, public participation in the process of memorialisation is still invariably fraught and partial; collective memory is rendered problematic by the conflicting emotions evoked at, and by, stigmatised sites.

Though contested, collective memory embedded in the built environment can be resilient. While sites can be signposted in an attempt to perceptually re-place the viewer in relation to past uses, there is also resilience in memory when the built environment is reused and transformed. While rebranding by new occupants might lend authority to different readings of places, reflecting attempts to reassign or control memory (Gough, 2004), buildings can be living entities inasmuch as their continued presence (whether intact or 'recycled') can speak to a prior institutional or ideological landscape. This latter observation of the power of 'bricks and mortar' to evoke memories is particularly salient in the case of re-purposing for educational activities because of the potential fit between asylum site plans and buildings and the needs of educational institutions.

For our purposes, we distinguish between the physical memorialisation of former uses of asylum sites and their remembrance as preserved through writing and other texts. Both memorialisation and remembrance can take on a range of forms and can be mutually influential: instances of remembrance may be promoted by memorialisation; and (perhaps more surprisingly) instances of remembrance can promote calls for memorialisation.

In a sense, memorialisation is concerned with (re)settling the past; attempting to offer some calm to put memories of troubling times to rest. This perspective aligns with Lowenthal's (1985) view that yearning for a more settled past is a key prompt for heritage preservation. However, just as guided remembrance through memorialisation can direct the 'tourist gaze' (Urry, 1990), then its opposite – 'strategic forgetting' – can be attempted through consciously minimising reference to past uses and events in publicity and signage. It is this latter process that we contend is a necessary undertaking in the quest to minimise the stigma associated with sites that have been vilified as outdated and undesirable. We now turn to the stories of our two primary case study sites, Lakeshore (Toronto) and Carrington/Oakley (Auckland) as a basis for investigating this process of strategic forgetting.

The Journey from Asylum to Higher Education

> ... a rundown relic of the past has been renewed and taken over by about 4,000
> college students with a great future ahead of them (*Globe and Mail*, 26/01/08).

> From the shadows of the past, Unitec has now established this place as a centre
> of learning which builds pathways to tomorrow (Unitec, 1994).

In this section, we recount the narrative of transition experienced by two facilities, the former Mimico Asylum, later Lakeshore Psychiatric Hospital, in Toronto Canada, and the formerly adjacent Carrington and Oakley Hospitals, in Auckland, New Zealand (Henceforth we refer to these facilities as, respectively, Lakeshore and Carrington). The case study approach allows us to unpack the rich detail of the transition from asylum to educational use. It also enables us to draw out the extent to which stigma is transcended or obscured and the techniques that are deployed to this end. At the outset we recognise that the re-use of asylum sites for higher education purposes is by no means new. In the UK, for example, the development and expansion of the University of Leicester took place on the site of the old Leicester and Rutland County Asylum (University of Leicester, 2014) which was closed upon the construction of a new and larger asylum in the years immediately preceding the First World War.

In the late 20th century, an aggressive programme of deinstitutionalisation leading to asylum closure coincided in many countries, with an expansion of higher education driven by population growth and aspirations to widen participation in the 'knowledge economy'. This convergence of intent created opportunities whereby one institutional use could be swapped for another: sites that had been attempting to heal the minds of a few could be reoriented to educating the minds of many. Again, the UK offers cases: Glenside Hospital, Bristol; Storthes Hall, Huddersfield; and Crichton Royal, Dumfries. Glenside is currently a campus of the University of the West of England, having seen a transition to what became that institution when nurse training moved into higher education in the late 1980s. Storthes Hall is now a residential complex for the University of Huddersfield, catering for significantly expanded student numbers. Crichton Royal became home to the Crichton Campus of Glasgow University and the University of the West of Scotland, offering higher education opportunities to the otherwise under-served south-west of Scotland.

Given the pervasiveness of the asylum model in the English-speaking world, such examples are not limited to the UK.

At both Lakeshore and Carrington, as well as at Challinor in Queensland, Australia, we had key informants on site who were able speak to the transition processes that occurred in each case. Both were developments inspired by philosophy and practice imported from the UK and the USA respectively. They were bold sets of large-scale buildings constructed for the long term and set in extensive and attractive estates. At Lakeshore, the (former) Mimico Lunatic

Asylum was designed by a leading provincial architect, Kivas Tully, between 1889 and 1894 as a grouping of two storey cottages on a 22.3 hectare core site. This design and style reflected a rejection of the monolithic structures that had hitherto characterised asylum architecture in Canada (Taylor Hazell Architects Ltd., 2007). The original plan called for a steady state population of 400 patients, but by 1903 there were 600, at which time 13.4 hectares of adjacent working farmland were added (Paine, 1997). In the our account we focus on the original, core site.

Prior to the mid-1860s, Auckland's asylum had been located in the elevated inner city neighbourhood of Grafton, site of the current Auckland City Hospital. This location was soon considered too close to the city and too small for the needs of a growing colonial city. The building of a replacement facility adjacent to Oakley Creek, approximately 7 km west of central Auckland began in 1865, and proceeded in an *ad hoc* fashion until 1915 (Avondale-Waterview Historical Society Incorporated, 2002). Using plans drawn up in England and bricks produced on the site, the outcome was "one of the largest public buildings in the colony at the time of construction" (New Zealand Historic Places Trust, 2009). In subsequent years the hospital site experienced fire, reconstruction and a number of additions in keeping with the neoclassical design of the original main block.

Acquisition for Educational Use

The stories of acquisition for educational re-use differ between Lakeshore and Carrington. Although the closure of Lakeshore Psychiatric Hospital was announced in September, 1979, the Ontario Ministry of Health did not declare the facility surplus to requirements until 1983 (*Toronto Star*, 7/10/86). In contrast, Carrington Hospital's closure was announced in July 1992 and a decanting occurred quickly with the hospital being emptied of patients by August the same year.

Humber College, which had a facility on an adjacent site, became a protagonist in the debate concerning re-use of the Lakeshore site in 1988 (*Toronto Star*, 2/12/88), offering a proposal to re-use the core buildings of Kivas Tully's quadrangle as a campus. However, the presence of high-density housing as a component of the overall development plan guaranteed continued public opposition in the local area of Etobicoke, as indicated by the headlines "Redevelopment projects anger Etobicoke groups" (*Globe and Mail*, 31/05/90) and "Proposed development in park protested" (*Globe and Mail*, 14/09/92). The word 'park' in the latter headline is important in interpreting the implied opposition; to local residents the asylum site was *both* a set of buildings imbued with heritage values *and* an uncommon example of an expanse of urban green space including highly sought-after public lake access. Opponents to the construction of housing on the site eventually won out with the news that : "After 15 years of debate, the fate of the Lakeshore Psychiatric Hospital grounds in Etobicoke was decided yesterday" (*Toronto Star*, 31/03/94). There would be a college and high school on the site, but no housing, and the dominant use would be public open space. We can read this outcome as a re-assertion of community rights regarding open space over government interests in accommodating diverse social and economic priorities in housing. Drawing on provincial funding, Humber College opened the first four renovated buildings as teaching and learning spaces in 1995 (Anon N.D.a). The renovations of additional buildings continued to keep pace with enrolment growth on what is now referred to as the college's Lakeshore Campus. A decade after the first arrival of students Humber College's new campus was described as "an absolute delight, using the renovated residences of the former psychiatric hospital" (*Toronto Star*, 15/10/04). The campus is now a dominant institutional presence within the community with its refurbished buildings and bold signage declaring the new mission of the Lakeshore quadrangle (Figure 5.1).

Figure 5.1   Lakeshore, Toronto, Canada: Humber Business School

In the case of Carrington in Auckland, an initial 19.5 hectares of land and hospital buildings was bought by the then Carrington Polytechnic for just under $5 million NZD from (the former) Auckland Area Health Board (*New Zealand Herald*, 13/2/93). The Polytechnic, later rebranded as Unitec, was an obvious candidate to purchase the remainder of the property. A planned $6 million NZD of internal revenues were to be spent in renovating the former hospital. According to the city's newspaper: "Cells that had been used to hold severely deranged patients could be easily transformed into officers for tutors" and representatives of the Auckland City Council and the New Zealand Historic Places Trust were reported to be "pleased with the polytechnic's approach" (*New Zealand Herald*, 13/2/93).

By 1994 the *New Zealand Herald* was able to report "That lovely 1896 building with its fretwork and veranda posts has been transformed into an airy student services and course information centre" (8/10/94). The Faculty of Architecture and Design became the principal occupant of the main quintessentially asylum building at Unitec. With growth however, the campus has also expanded into other former hospital buildings. After the purchase of the remainder of the adjacent hospital land and buildings, Unitec campus now spreads over 55 hectares (Unitec N.D.c) (Figure 5.2).

Figure 5.2   Carrington Hospital, Auckland, New Zealand, now Unitec

Memorialisation and Remembrance

In the Lakeshore case, the asylum is implicitly memorialised at a physical level through the very bricks and mortar of Kivas Tully's quadrangle. According to the restoration project's lead architects " ... the original architectural concept and grouping formulated as the Mimico Asylum has remained more or less intact ..." (Taylor Hazell Architects Ltd., 2007, p. 3). In much of the discussion of the re-development of the site for educational uses, emphasis was placed on the architectural merits of the complex in the first instance rather than on its past use. As has been the case with many other instances of architecturally-motivated 'preservation' in Canada, it is also only the exterior of buildings that have been protected with interiors remodelled so they are fit-for-purpose. According to publicity: "The cottages have been restored to their original architectural integrity, with interiors converted into state-of-the-art classrooms" (Humber Institute of Technology and Advanced Learning and Assembly Hall City of Toronto, N.D). The interior layout and fittings of these buildings strongly evoked their former use and with their removal, this tangible association has been lost. In terms of evidence of former use, little apart from the exterior appearance of the buildings themselves remains. Only one sign, for COTT 3 (Cottage 3), at the rear of one of the un-renovated residential buildings signals this era (Figure 5.3).

Figure 5.3   Cottage 3, Lakeshore, Toronto, Canada: Humber College Campus

In contrast to the absence of specific memorialisation within what is now the Humber College quadrangle, the nearby Assembly Hall offers a formal memorialisation of the past use of the site. Paradoxically, and perhaps pertinently for our analysis, the hall is owned not by the College, but rather by the City of Toronto. In and near the Hall, there are specific references to the lives of the one-time residents. The previous use of this building and its construction by patients in 1897 is openly declared in the Assembly Hall itself. However, it is in the 'Third Garden' adjacent to the Hall that the lives of residents are recounted in an art work "consisting of five cast iron seating forms, eight concrete pavers inlaid with cast bronze text, and landscaping ... .[that] provides psychological space and creates a memorial" (Toronto Culture, N.D.) The paving stone installation employs the written words of residents drawn from across the history of the asylum in order to metaphorically re-populate the erased space of the asylum (Figure 5.4). Some examples are:

June 30 1932 "Went out with the other patients to cut and rake the grass, pick strawberries from the garden near the Assembly Hall. Got a good workout. Come in good and hungry."

October 11, 1958 "The gardens have been taken away from us. We can't work them anymore. Something about cheap labour."

September 1, 1979 "They're closing the hospital soon...We're being taken to a place we don't know. Some cried."

Figure 5.4   Third garden paving stones on site of the former Lakeshore Hospital, Toronto, Canada

Juxtaposed with this rare instance of formal memorialisation is a shifting patchwork of remembrance. While such remembrance includes urban legend – "The campus is rumoured to be haunted by ghosts, lurking in the maze of tunnels that connect the various buildings in the complex" (*Toronto Star*, 20/09/03) – it also embraces formal, mainstream remembrance. Examples range from "A Summer's Walk through Mimico Asylum" in August, 2007 sponsored by the Toronto Architectural Conservancy (Micallef, 2007) to the self-guided discovery trail encouraged by the freely-available booklet *The Lakeshore Grounds: A Community of Learning, Recreation, Creativity, Caring and Stewardship* (Humber Institute of Technology and Advanced Learning and Assembly Hall City of Toronto, N.D). Again, however, it is the material heritage – the buildings and grounds – that are celebrated rather than former residents themselves.

Humber College's Lakeshore Campus is clearly now more than a " ... a spooky group of red-brick buildings ..." (*Toronto Star*, 30/10/00). Although there are minimal tangible reminders of the previous uses of classroom buildings, the site's history is not obscured from those who are curious and wander into the Assembly Hall or contemplate the Third Garden. Further, evidence of the past is available to those who turn to the internet for illumination. Websites such as The Former Lakeshore Psychiatric Hospital (Barc, 2009) offer extensive but sometimes uncorroborated information about past uses of the site.

In the case of Carrington we can observe an analogous transformation into what has been rather optimistically termed a 'living learning community', a phrase in which we speculate that the use of the words 'living' and 'learning' signal a shift from associations with the past:

> From a distance there's still something a little creepy about the Victorian brick façade of the old Carrington Hospital. Peering with suspicion through the trees at the forbidding walls of Oakley/Carrington has been part of the Auckland psyche for decades ... a place where the "loonies" were locked up out of sight .... But take a look at it now. There is something faintly bizarre about the sight of a young woman student sitting in the lunchtime sunshine on the doorstep of what used to be the notorious Oakley M3 secure ward for hardened, disturbed criminals (*New Zealand Herald*, 8/10/94)

From the outset, the new owners were faced with a tension between the options of emphasising remembering and forgetting. On the one hand, the new educational purpose was to be endorsed with a forward-looking philosophy but, on the other, historical building protection/designation meant there were severe limits on erasing the collective memory of former use. According to the consultant architects who were commissioned to develop a conservation plan:

> While the place may have unhappy associations for some, and while it may be seen to represent a part of the history of Auckland which it has been convenient to overlook, the building is a noticeable work of architecture and a valuable community resource (Salmond Architects, 1993, p. 1).

The architects went on to identify the rare 'fit' between asylum and educational uses that could fortuitously be achieved on the Carrington site:

> It is rare that a new use can be found for a large historic building which can so readily make use of its intrinsic architectural and spatial characteristics. In this case, the physical needs of the Polytechnic can be fulfilled with remarkable facility in the old hospital building with great economy of means. The result promises to have immense community value and very great corresponding educational value for the design-oriented departments which will be housed in the building (Salmond Architects, 1993, p. 2).

Shortly after its acquisition of the buildings, a five-page advertising section for Unitec that appeared in an Auckland community newspaper highlighted only pleasing aspects of the former asylum site – its "fine park like environment", the "spacious" campus, and the "historical buildings" (Western Leader, 20/1/94). The advertising feature repeatedly included the statement "formerly Carrington Polytechnic", as if to create distance from the name Carrington An emphasis was placed on the advantages of Unitec having a 'real' campus (a tacit point of distinction from the University of Auckland and Auckland University of Technology, both of which occupy central city sites):

> Unitec Institute of Technology offers its students the full service and facilities of a real campus. Students are now able to enjoy first class teaching facilities and resources in a total campus experience. Even better, you will be studying in a fine, park like environment only eight minutes from the centre of Auckland City ... . The campus is spacious, the buildings are a pleasing mix of historical and modern – many of them custom designed. (Unitec, 1994)

Later that year, the site was rededicated by the then Prime Minister Jim Bolger, who unveiled a plaque that, "acknowledges the past and looks forward to the future" (Unitec, 1994). This is one of only two formal instances of memorialisation of the former use. The other states:

> This building was originally established as part of a psychiatric hospital. From 1896 to 1989 thousands of New Zealanders came here through a system of health care that deeply affected their lives. For some it was a place of confinement, for others a refuge. From the shadows of the past, Unitec is building pathways to tomorrow and has established this place as part of a centre of learning. We acknowledge the previous residents of this building with respect and honour the students who, from this same place, will shape New Zealand's future.

Informally there is another more poignant legacy of former Carrington patients – names and patterns etched into the surface of some of the gold-coloured bricks in an exercise yard (*New Zealand Herald*, 8/10/94). Recently a further memorial has been added in the form of a carved *pou* (internal pillar) within the *whare nui* (Māori meeting house) on campus. This carving honours a particularly well-loved patient and artist as well as and offering recognition to the often traumatic impact of asylum care on Māori (Kingi, 2005; Baxter, 2008).

The current Unitec website contains an historical account that emphasises only educational developments on the site (Unitec, N.D.b). While the sub-page on the Mount Albert (Carrington) Campus (Unitec, N.D.a) links the former 'Carrington' name to the bordering Carrington Road (itself named in honour of surveyor general Frederick Carrington) there is no mention of the former Carrington asylum. In 2010, previous inspections of the website revealed references to Unitec's historical association with "hospital land and buildings" but in 2014 the only remaining reference to the asylum use on the site is reflective of the popular culture view of the asylum (see Chapter 7): "Gemma Duncan admits she didn't know much about Unitec other than playing a zombie for a film production, wandering the halls of the old Carrington Hospital, now part of Unitec's Mt. Albert Campus" (Unitec Website, accessed 19 March, 2014). This choice arguably amounts to a case of strategic forgetting, rather than remembrance. Beyond the official image of Unitec, this forgetting is less evident. A link is made on the website between one of Unitec's teaching strengths (building design) and its heritage (the former hospital buildings): "As part of the Faculty's commitment to architecture, construction and design, the building's unique character will be preserved and cherished" (Unitec,

N.D. c). This link was further reinforced through certain hospital buildings being designated under the Historic Places Act (1993), which is applied to places of special or outstanding historical or cultural heritage, significance or value. This designation influenced the crafting of 'Guidelines for the take-over of the Carrington Psychiatric Hospital' (Unitec, N.D. d). These imperatives state that "the planning and use of the former Carrington Hospital building should take account of its heritage significance". They also suggest that the history of the building could be incorporated into the teaching programme in the Faculty of Architecture and Design. This objective appears to have been honoured with the interior of the building housing displays that commemorate the former use, and the interiors of some of the buildings remaining largely faithful to their original design. In the former male wing known as Oakley Hospital small rooms that were little more than cells have been converted to art studios and the corridors feature corner observation ports with three narrow panes of glass through which the warders could check up on the occupants within. These portals remain clearly visible and evoke a history of confinement (Figure 5.5).

Figure 5.5   Former Asylum Cell converted to design studio, Unitec, Auckland, New Zealand

A further way in which memorialisation prevails is the sheer scale of the campus and the assortment of many under-used buildings some of which remain in poor repair. The remarks of an early commentator hold true: "While the restoration of the main building is nearly complete, a closer inspection of the newly extended campus reveals odd corners and buildings which remain firmly stamped with the marks of a psychiatric institution" (*New Zealand Herald*, 8/10/94). Translating the asylum to educational use has required on-going investment that has not always been possible given the vicissitudes of higher education funding.

Unlike Lakeshore, where a decade-long hiatus between asylum and educational uses created a 'temporal buffer', there was a clear overlap at the Carrington site where the transition process saw a co-location of patients and students not only in juxtaposed buildings, but even within partitioned sections of the main building. Further, the closure was so fast that many patient records remained on site and were later delivered to the hospital board offices by Unitec staff. This interleaving of institutional spaces meant that long-term Unitec staff members have a collective memory of the psychiatric facility and this has informally kept vivid remembrance of past use alive. Although the hospital closed in 1992, psychiatric/educational co-location has also continued into the present with the Mason Clinc which houses forensic psychiatric patients located at the rear of the Unitec grounds (see Chapter 4). This co-location may, in part, explain the relative dearth of formal memorialisation.

Educational Re-use for Asylums: Opportunities and Constraints

Having recounted the narratives associated with two key examples of educational re-use of psychiatric asylums, we now return to reflect on the themes underpinning our core concerns in this chapter: remembrance and memorialisation of past psychiatric uses. As we have contended earlier, as the process of deinstitutionalisation unfolded, there was an understandable focus that rested primarily on people rather than buildings. The substantial infrastructure that progressively became surplus to newly expressed needs stood almost forgotten as health planners turned attention to the challenges associated with gaining community acceptance of a new modality of care (Dear and Taylor, 1982; Philo, 2000; Wolch and Philo, 2000). This process proceeded at different rates and in different ways. In Canada it resulted in buildings and land lying vacant for a significant period of time during which debates over future use were able to flourish and ultimately, at least in part, frame the extent of educational reuse at Lakeshore. In New Zealand, the interleaving of psychiatric and educational uses at Carrington arguably reflects what we might call 'drift closure'. Asylum use tapered off slowly and educational takeover overlapped. This situation was symptomatic of the ambiguous and almost 'non-policy' environment that prevailed in New Zealand mental health care as the country belatedly shifted from an institutional to a community model of care (Hall and Joseph, 1988).

The unfolding of deinstitutionalisation policy also provides a general underpinning to remembrance and memorialisation. First, in the case of Carrington, the ambiguous character of the closure process ensured that the new users of the former asylum had an active engagement with the earlier use and an experiential basis for remembrance. More significantly, despite being stigmatised, the asylum sites in both countries had a strong symbolic presence in their local communities. This presence had involved a role as a community resource expressed in ways such as informal access rights to the asylum grounds. Deinstitutionalisation challenged this position with surplus land and buildings potentially being regarded as an opportunity to reduce public debt through a transfer to private ownership. Such a privatised future would diminish or remove the community use of the former asylum site. Educational re-use enabled redeployment within the public sector and the notion of the site as a community resource was maintained.

In this trajectory from health care to education we can discern parallels with what has been called transinstitutionalisation (Gleeson and Kearns, 2001; Hudson, 1991; Moon, 2000). This term has generally been used to refer to the post asylum movement of clients through the community care system, to the shuttling and churning of individuals through the 'revolving doors' of the prison and health care systems, and to continuing institutionalisation within the context of community care. Here we redirect the term to refer to the movement, not of clients, but of the tenure of buildings and estates. It is the sites themselves that have transinstitutionalised. Part of the rationale for this transinstitutionalisation involves retention of a community resource. Another relevant observation is a concordance between educational and asylum space needs. Here we recall the cell-studio spaces at Carrington/Unitec and the educational resonances of Kivas Tully's quadrangle at Lakeshore/Humber. Thirdly, there is the sense in which both education and mental health care share an institutional focus on the mind.

The urban fringe location and physical accessibility of both former asylum sites is a further theme to note. These attributes might equally relate to conversion to housing or to educational uses. In some instances, accessibility to other amenities (including employment) and escalating land values have been the imperative for rapid recycling of urban hospital sites (Chaplin and Peters, 2003). In other instances, particularly in more remote settings, this recycling can be contested and more drawn out (Cornish, 1997; Joseph et al., 2009). In the Lakeshore and Carrington cases the (large) size of sites, their location (inner suburbs of metropolitan centres), and their buildings (amenable to refurbishment) rendered them excellent prospects for educational re-use. Moreover, the adjacency of pre-existing buildings of the higher education bodies that took over the sites also contributed to their amenability to conversion. Finally, in both cases, the acquisition of the former psychiatric hospital site and buildings was predicated on a bold and aggressive move into degree level educational provision; it was success in this venture that produced both capital and operating funding for the ongoing renovation of the Lakeshore site.

Once acquired, our two case study sites provided challenges regarding the management of 'heritage', not unlike other 'brown field' urban uses such as converted loft apartments in former warehouses and the tourist marketing of 'heritage' industrial buildings as discussed in Chapter 2. At issue is the extent to which previous use should be memorialised or whether the emphasis should be on the new use and a consequent suppression of memories. Our case studies indicate limited engagement with heritage themes and a similar situation is evident in the case of reuse for housing (see Chapter 6) (Weiner, 2004). Heritage designations facilitate the preservation of grounds and building facades but little more. The heritage literature leads us to contend that the absence of more extensive memorialisation and remembrance can be linked to the stigmatised past of the asylum sites. It was the usage, the practices of asylum care, that was stigmatised and not the sites themselves. The ending of asylum use left (only) an asylum site; over time this site emerged from the shadow of the stigmatising use and could be re-purposed for a new future. We observe that the educational future is sufficiently different to the preceding asylum use that it lacks contiguity with the past, unlike, for example, industrial sites becoming museums, and thus memorialisation is limited. The installation/existence of a limited number of memorials at Carrington and Lakeshore speaks, in words and symbols, to the lives of those who lived and worked in the former asylums. Educational re-use constitutes a radical departure from the asylum, and the question of how much and in what style such memorialisation should occur is paramount. We contend that relict uses can, however, still speak to the past, particularly when educational budget constraints limit the extent of renovation. At both Carrington and Lakeshore, corners of the sites remain visually run-down and quintessentially asylum-like in appearance, in essence, 'spectral signs' (Edensor, 2005b) (see Chapter 7). This limited, fragmented and fleeting evidence of the past can, following DeSilvey (2006; 2007), be seen as a spark to memories of the former use. Its juxtaposition with the new serves as a frequent reminder of a not-too-distant past as well as a catalyst for more exotic remembrance. By way of example, an issue of the Unitec student magazine in 2006 featured a five page photo-essay headlined 'Looneytec', which included accounts – both historical and apocryphal – of events and practices within the buildings of the former asylum (McKechnie, 2004). Intriguingly such voyeuristic aspects of remembrance also manifest themselves on the internet in reports of occasionally clandestine 'exploration' of the otherwise forgotten corners of re-purposed sites. Such exploration is of course usually associated with derelict sites, and such incursions will be considered in detail in Chapter 7.

In a general sense our case studies suggest that there appears to be a limited collective will to materialise remembrance in an overt manner. The memories (and traumas) are personal, indeed, often also associated with shame. Hence our finding of plaques that are modest in scope and inscription is perhaps not unexpected. Lakeshore's Third Garden captures this personalised memorialisation in a way that resonates with the sorts of interpretations that have been made of memorial benches (Wylie, 2009) and trees (Cloke and Pawson, 2008): it carries memories of the past into the present day and serves as a reminder of past presences. On the collective side, the *marae* (Māori meeting house) at Unitec speaks to perhaps the most clearly articulated set of collective memories: the place of the asylum in recent Māori history. For here, among the intricately carved wooden *pou* (internal pillars), there is one which depicts a Māori psychiatric patient reaching out from behind the bars of a secure window set within the characteristic brick walls of the former Carrington Hospital.

To round out our discussion, we now turn to two supplementary case studies. They differ from the primary case studies in two important case-specific ways. The Kalamazoo example, represents an early stage of transition, in which the psychiatric institution remains open, but diminished in size and dwarfed by the adjacent and rapidly expanding university campus. In contrast, the Challinor example represents an instance in which the psychiatric institution has closed, but several un-refurbished buildings remain alongside those re-purposed for the new educational use.

The Michigan Asylum for the Insane opened in Kalamazoo in 1859. Consistent with developments elsewhere in the US, its name was changed to the Kalamazoo State Hospital in 1911 and to the Kalamazoo Regional Psychiatric Hospital in 1978. The regional designation was dropped in 1995, but the hospital remains a major provider of mental health services in western Michigan. From a peak of 3,500 patients, the resident population had dropped to around 300 in 2012 (Kalamazoo State Hospital, 2014), although the capacity of its large and impressive buildings remains significantly higher. Kalamazoo is relevant in this chapter because it exemplifies the ongoing penchant of governments to retain land and to seek synergies between excess capacity in one sector and need in another. In this particular instance, the State of Michigan has already transferred most of the former hospital estate to Western Michigan University. The latter has removed a prominent former hospital building and re-developed land that previously existed as a buffer between itself and the hospital. Most of the former hospital land (with or without buildings) along the main access drive to the hospital is now controlled by the University (Kalamazoo State Hospital, 2014), Characteristically, in echoes of other transitutionalisations involving the education sector, the developments at Kalamazoo represent something of a strategic transition in that the new university buildings have a focus on health care and nursing. One cannot help but speculate that the surviving and grossly under-used hospital buildings will, sooner rather than later, be re-developed with the blessing of the State government as the focus of an expanded health care campus. In terms of memorialisation, signage and historical plaques related to the still-operating hospital are obvious cues to remembrance of a past in which the asylum was the major institution in the city. However, it is not clear whether these pointers to the past will remain if (or perhaps more likely, when) the university completes its acquisition of the site.

A rather different but equally transparent presentation of institutional history can be seen at the Ipswich Campus developed by the University of Queensland (but soon to be transferred to the University of Southern Queensland) which occupies the site and buildings of the former Challinor Centre (Challinor was the most recent of four institutional identities and names, the original being the Woogaroo Lunatic Asylum built in 1878). Compared to the rather muted acknowledgement of Unitec's institutional prehistory on its website, the small-scale Ipswich Campus offers those who are curious a virtual window into the past by means of its Library website (University of Queensland, 2014).Here, under the banner of 'Progress of an Institution', the site offers visitors the opportunity to learn of the six institutional eras and offers vignette-based subpages concerning administration, buildings, grounds and residents. The candid embrace of the campus's asylum past is striking, with the introduction to the site invoking the quest for community as a link between past and present uses: "From a place for society's less fortunate, it has become a place for those fortunate enough to receive a university education. It will always be a place about community" (University of Queensland, 2014). Notwithstanding this transparency, students are only variably aware of this asylum history (Rix, 2011 personal communication) which is nonetheless more materially evident than in other case study locations. On the campus, for instance, we noted not only plaques acknowledging previous use (Figure 5.6), but also one particularly recognising the abuse that occurred in a previous era. Elsewhere in the library, there is artwork by a former resident, and security bars remain on many windows of refurbished buildings. Further, abandoned but intact villas on the periphery of the site reference the recency of closure and re-purposing. For those keen to engage more closely with the campus history a journey through the past can be undertaken by means of a self-guided walk which takes the visitor to sites such as the original industrial plant where a restored large-scale boiler sits juxtaposed with computer workstations. Elsewhere, the asylum's theatre has been restored into a lecture space with the backstage area still set up as it was for performance days in the asylum era. This degree of retained reminders of the past can be explained by the firm heritage protection requirements in place at the time of purchase, along with the passion of some key players (notably the current Pro-Vice Chancellor in charge of the campus) to keep the past evident within the suite of current uses.

Collectively, the four examples of transinstutionalisation considered in this chapter underscore the opportunity for tertiary education facilities to take advantage of the availability of former asylum sites and buildings. Consistent with our discussion in Chapter 2, they also support the importance of policy in circumscribing opportunites at different points in time. They also point to the consistent importance of a key conjunction for responsible governments between the decline of one demand on land resources and the rise of another, both of which are associated with place identity and job creation. Nevertheless, within this strong structural momentum human agency, sometimes if not frequently hidden from view, has a clear role, especially in decisions concerning the memorialisation of the past use.

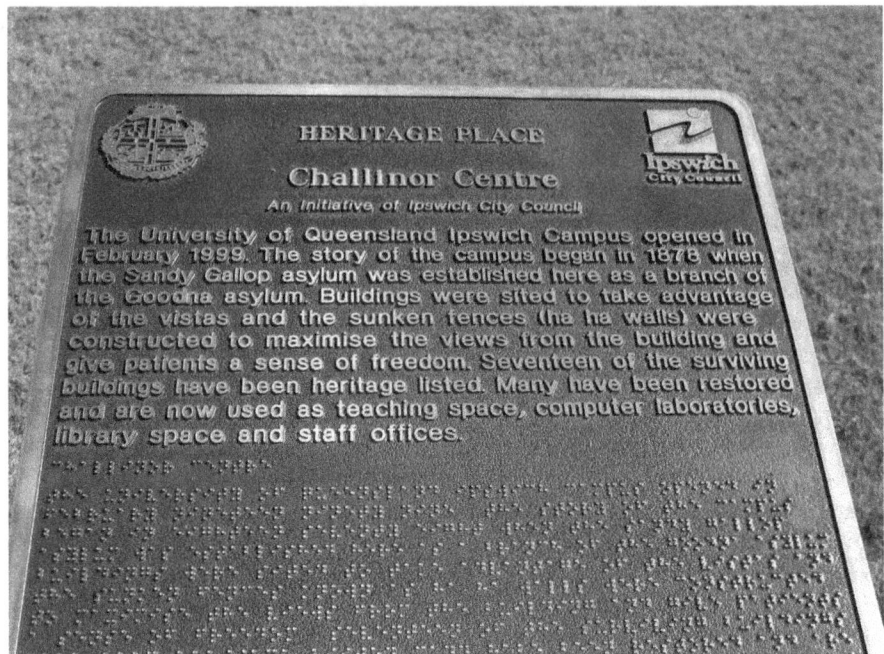

Figure 5.6　　Memorialisation at Challinor, Queensland, Australia

Re-use: The Past Meets Future Orientation

The understandings of present-day educationalists about the meaning of an inherited site differ substantially both from the memories of former psychiatric inmates and also from the understandings held by those involved in the immediate purchase and conversion of sites. As noted in Chapter 2, memories are not fixed and immutable (Landzelius, 2003) and remembering is re-membering: a putting-back-together of the past. In our case examples, this re-membering is inevitably partial, not only by virtue of contested memories clouded by differing notions of stigma, but also because only the buildings remain. The people, actions and performance of psychiatric care have gone. That which once made the asylum has been erased and we are left with sites haunted by absence. Into these spaces have flowed educational uses with recast futures disturbed at the margins by echoes of a receding, partially remembered, and potentially disturbing past.

　　Further insight into the relative absence of memorialisation may lie in the fact that new higher education uses at former hospital sites often remain in a formative state. The 'becomingness' (Pred, 1984; Rose, 2006) of the site as a place of higher education emphasises futures and new beginnings rather than stigmatised pasts. This emphasis was evident at all four sites we examined, especially in those early phases when evidence of the past was enshrined in relict buildings replete with the furnishings associated with the past use. A key task for the new owners of the former asylums is to incorporate these sites into their visioning of a future within highly competitive educational markets. The former use does not speak to the new future and thus tends not to be declared by the new occupants. Instead it needs to be discerned by the knowing observer. Some material imprints of the asylum past are sufficiently tenacious to reach into the educational present but they do so with the connivance and acquiescence of the new owners and reflect new realities and requirements, and these are subject to local regulation and the associated actions of local leaders charged with their interpretation. Specifically, the representation of the past in contemporary public spaces necessarily reflects current structures of power and organisational needs (Azaryahu and Foote, 2008; Foote and Azaryahu, 2007). Moreover, memorialisation is complicated with the passage of time (Burk, 2003). We speculate whether the maturing of higher education projects on former asylum sites will, in time, result in a coming-to-terms with the stigmatised past of the sites and greater memorialisation, or a complete and very strategic forgetting. Our case studies suggest that educational providers are faced with a dilemma: to attempt to silence the past or offer 'too much memory' and risk unsettling their own aspirations. Only in specific situations, such as the occasional art exhibitions on the theme of confinement at Unitec as well as its use of secure cells as artists' studios and the 'memory walks' at Lakeshore and Challinor do we see a vestigial engagement with the fundamental realities of past use.

We see rebranding as a key element in this strategic forgetting: the transition to, and maintenance of, a new institutional identity at both Carrington and Lakeshore, involves what we might also call 'selective remembering'. Publicity references the historical character of the buildings, but not their historical uses. At Carrington, it speaks of teaching spaces including "inspiring historic buildings" (Unitec, N.D.a) but subsequent advertisements show only modern buildings and the park-like setting. We see the new owners stressing the positive externalities associated with the grounds but avoiding visual imagery that might hint at the asylum origin. It seems that, in situations in which uses are transformed from a non-competitive to a competitive environment, considerations of memorialisation and remembrance are mediated by the imperative to re-brand a key asset and strategically forget the stigmatised asylum past. In the contemporary educational marketplace in which degree programmes in competing institutions may appear much of a likeness, the amenity-value of sites is of paramount importance in attracting students. To this end we see the site and design characteristics of the former psychiatric hospitals being broken down into their positive components but largely separated from the stigmatised use that gave rise to their initial formation. In both case studies we note a broad support for the transition between uses from surrounding communities whose residents see a guarantee of the preservation of green space that would not occur to the same extent if the former asylum became a residential development.

While we are conscious of the limitations that flow from the use of only a few exemplars of a complex process, we believe that our examination of selected stories reveals much about the journey from asylum to educational use. This re-use has been an expression of the recent expansion of tertiary education that has created an appetite for heritage buildings in what could easily be re-imagined as a campus setting by new owners, a process that Hoskins has called "turn(ing) a history of exclusion into a future of inclusion" (Hoskins, 2007, p. 437). To some extent at least, this willingness to at least partially embrace a troubled and stigmatised past could be seen as a deeply-engrained characteristic of transinstitutionalisation within the public sector. As will be seen in the next chapter, comparison with the public-to-private transition involved in re-purposing for housing certainly reinforces the foregoing observation.

Chapter 6
Re-imagining Psychiatric Asylum Spaces through Residential Redevelopment

In this chapter, we examine the recycling of asylum sites and (sometimes) buildings for residential purposes. As noted in Chapter 1, this is a common successor use noted in surveys such as that reported by Chaplin and Peters (2003) in the UK. Examples of re-development for residential use can be found in many, if not most, jurisdictions. In the United Kingdom, the former Exe Vale Hospital (originally the Devon County Pauper Lunatic Asylum) has been re-incarnated as Devington Park, a gated residential community (Franklin, 2002). In the United States, the iconic Danvers State Hospital in Massachusetts (originally the State Lunatic Hospital at Danvers) has been partially re-purposed as Avalon Danvers, a complex incorporating owner-occupied units on cleared parts of the former hospital site and rental units in the retained central portion of the Kirkbride building that once stood at the core of the hospital (http://ma-smartgrowth.org/wp-content/uploads/Danvers_Avalon.pdf). In New Zealand, the site of the former Ngawhatu psychiatric hospital near Nelson has been re-purposed as the Montebello housing development (Kearns et al., 2012). Thus, it is something of a surprise that the phenomenon has received only limited scholarly attention. As noted elsewhere in the book, this neglect might reflect the ambiguity surrounding these sites. They were, at once, architecturally distinguished complexes of buildings in extensive landscaped settings – secluded, separate and contained – and liminal spaces between the old and the new; zones of ambiguity stigmatised by the shadow of their former use.

> Sweetheart deals with developers saw the great Victorian [asylum] parks, with their dark histories, their infamous architecture, pillaged, revamped and repackaged … (Sinclair, 2002, p. 166)

> … The map of the site was dotted with strange names – Connolly Mews, Chapel Square, The Piazza – that owed little to its previous existence. How much more appropriate I thought, if they had given them names like Lobotomy Square and Electroconvulsive Court … (Bryson, 1997, p. 83)

The above quotes serve as an introduction to the themes we develop. They tell of sites with complex histories and their potential as exclusive housing developments. They also tell of the erasure of the memory of former use. Intriguing questions are raised. For instance, how, if at all, is the former asylum use remembered on sites and in buildings redeveloped for housing? How are the concepts of strategic forgetting and selective remembrance introduced in Chapter 2 manifested in the quest to build a new use and future on the sites? To what extent are such processes and outcomes normative and how do they relate to broader trends in the recycling of 'brownfield' sites for housing? As in the preceding chapters, in approaching these themes we deploy a case study approach. In this instance, we present two examples from the United Kingdom, Graylingwell and Knowle, and one from New Zealand, Sunnyside. We describe how the peripheral (and attractively landscaped) sites of many asylums have led to their appeal for housing and transformation into (paradoxically) often gated communities, noting the use of the concepts of strategic forgetting and selective remembrance in the plans and marketing tactics adopted by developers. In probing the role of government in shaping re-development, we pay particular attention to the impact of heritage designation and land use controls. We connect our discussion of what might seem to be particular and idiosyncratic into the mainstream of discourse on urbanism and land use planning by considering parallels and contrasts with the re-use of former military and industrial sites and infrastructure.

The remainder of the chapter is organised in four major sections. We turn first to an examination of context, proceeding from a review of research on the re-use of former psychiatric asylums for residential purposes to a consideration of overarching policy frameworks. Second, drawing on the three case studies we assess the extent to which asylum use is memorialised, strategically forgotten or selectively remembered in the successor residential developments. Third, we distil implications from the case study narratives for the theorising of processes by which former psychiatric hospitals are re-integrated into the built environment as housing developments. In a concluding section we consider the implications of the observed interaction of heritage conservation policies and residential re-use and draw out parallels with other types of brownfield re-development.

Contextual Background

Built Environment: The Re-use of Psychiatric Hospitals for Residential Purposes

> These spacious sites are often in excellent locations on high ground, with fine views, on the edge of (or now sometimes surrounded by) towns and cities ... the demand and value for these sites would be enormous (Lowin et al., 1998, p. 130).

Speaking to the results of a comprehensive survey conducted in 1996 of 206 large (> 100 bed) UK psychiatric and learning disability hospitals, Lowin et al. (1998, p. 129) reported that 'residential purposes' loomed large in planned use for vacant sites, but noted that "obtaining planning permission for the sites is a lengthy and sometimes impossible process. Public opposition to site development is often fierce. Further, many buildings are listed or have conservation orders on them, and sites are often on green belt land" (Lowin et al., 1998, p. 129). A few years later and despite the barriers identified above, the emergence of residential development as a favoured re-use of former psychiatric hospital in the UK was coming into clearer focus. Chaplin and Peters (2003) surveyed 71 hospitals in six areas of England to determine the proportion of hospitals still open and the fate of those that had closed. Preserved buildings were reported on more than a third of the repurposed sites, often as part of 'luxury' housing developments. Indeed, the authors reported than six developments were "entirely private with no public access, often guarded by security guards" (Chaplin and Peters, 2003, p. 227).

The remaining papers in the small literature on the afterlife of the psychiatric asylum provide an indication of similar processes in countries beyond the United Kingdom. In an examination of three cases in New Zealand (Carrington/ Oakley and Kingseat in Auckland and Seaview in Hokitika) and one in Canada (Lakeshore in Toronto), Joseph et al. (2009) and Kearns et al. (2010) report that re-development for housing was seriously considered or pursued in all but one case, that of Carrington/Oakley. At Seaview and Kingseat, the current owners still aspire to re-develop at least a portion of the sites for housing (Kearns et al., 2012). In the case of Lakeshore, housing was a prominent and contentious component of the re-development plan for six years, and was only set aside after the surrounding community had vigorously (and effectively) asserted their rights of access to one of the only remaining extensive green spaces in the city with access to the Lake Ontario shoreline. The Lakeshore outcome reminds us of the merits of incorporating both the external as well as internal characteristics of sites into our case studies, a lesson also noted in Chapter 4 with respect to the encroachment of housing at St James' Hospital, Portsmouth. External characteristics include the political as well as socio-geographic circumstances and processes that selectively constrain and/or promote both particular types of re-use and the preservation of cultural landscape.

Policy Frameworks: Stigma and the Cultural Landscapes

Between active asylum use and the opening of the housing developments described by Chaplin and Peters (2003) lies the question of stigma (Goffman, 1961; 1963) and our particular interest in how it is (or is not) addressed through mechanisms of memorialisation and remembrance. The stigma associated with the psychiatric asylum – the almost universal characterisation of it as an inhumane and outdated treatment modality – was, as noted in Chapters 1 and 2, accentuated in debates that occurred during the years immediately prior to closure and has remained salient in debates over re-use. Thus it is not surprising that, as a built form, the asylum can be seen to have an "ambivalent quality" … "symbolic of fear and oppression, but architecturally impressive monuments" (Franklin, 2002, p. 174). In the UK, local health authorities, which were almost always charged with the disposal of particular asylum buildings and sites in the 1980s and 1990s, were, in the absence of guidance from senior levels of government, free to chart their own course through this ambivalence. As Franklin (2002, p. 174) notes, " … in some cases, they have demolished the buildings, as a cleared site is potentially more profitable. Others have been left derelict for several years whilst their future has been debated". Gradually, however, in the UK there emerged pressure on two fronts concerning a more coherent approach to re-use. First, and not specific to residential re-development, there were increasingly urgent calls for the conservation of a distinct part of the built environment (SAVE, 1995). Second, and specific to residential re-use, there was recognition at the highest levels of government that, as much as possible, demands for urban growth should be met from 'brownfield' sites (Franklin, 2002). In this latter context, the availability of sites tainted (merely) by the memory of past use, as opposed to contamination by industrial waste, might be expected to be attractive, especially when their often generous endowment with surrounding green space is taken into account. The potential returns on re-development of brownfield sites for housing – estimated in the late 1990s to be up to five times more profitable than repurposing for institutional use (Lowin et al., 1998) – provided a strong fiscal rationale for promoting and implementing policy linking brownfield site development and heritage conservation priorities.

In England, the strategy of using the profits from housing development to finance heritage conservation began to be actively promoted by English Partnerships (now part of the Homes and Communities Agency) in 2005, and was subsequently concretised through their acquisition of 96 former hospitals (including a number of former asylums). With a commitment to incorporate affordable housing as well as preserve heritage and promote sustainability, this undertaking constituted a critical framework in the UK for decision-making on re-use, replete with implications for memorialisation and remembrance. Policy frameworks were less developed elsewhere, including in New Zealand, as will become evident later in this chapter. In understanding how such a policy emerged in the UK when it did, Franklin (2002, p. 175) suggests that the growing interest in re-developing asylum sites for housing in the late 1990s could be attributed in part to the recovery of the housing market and to the positive example provided by several successful conversions in key locations. She also suggests that another important factor might be at work – " … the elapse of that necessary period of time during which the stigmatising association of the pre-existing use can begin to evaporate and be replaced by acceptance … ".

In terms of memorialisation and remembrance, Chaplin and Peters (2003, p. 228) report that while property developers often deployed adjectives in their advertising – such as 'seclusion' and 'sanctuary' – that could be applied to the predecessor asylum uses, they very rarely made reference to those former psychiatric uses, "possibly reflecting the stigma of their former existence". According to the authors, "paradoxically, asylum can now be bought in an ideal self-contained community, with security to keep society out" (Chaplin and Peters, 2003, p. 228). Our earlier quote from Bill Bryson's UK travelogue provides an ironic comment on this situation in respect of a former private asylum – the Royal Holloway Sanatorium. Similarly, Franklin (2002) reports that the former Exe Vale Hospital (built as the Devon County Pauper Lunatic Asylum, surely a double-dose of stigma) was re-packaged as Devington Park, with mention of the previous use conspicuously absent in marketing material. As the developer of Devington Park put it: "God no – steer clear of that, 99% of people don't want to live in a mental hospital" (Franklin, 2002, p. 177). Indeed, it seems that in the UK "the former asylums have been reappraised, not as containers of madness, but as unique works of architecture" (Franklin, 2002, p. 183). Stripped of their original name and identity, and set as they are within extensive parklands, the casual observer could be forgiven for mistaking a re-cycled asylum for the buildings and grounds of a refurbished stately home. In a later commentary on Devington Park, Bowden (2012, p. 120) postulates a clash between the gentrification pursued by developers and the psychogeographies of those with links to the former asylum, with security cast as a potential good that constitutes "a new confinement, keeping people out as opposed to in". We take from these observations a need to expand upon the concept of strategic forgetting deployed in Chapter 5. While the latter is very evident in accounts of the re-development of former psychiatric asylum sites for housing such as those provided by Franklin (2002), it seems that there is sometimes a companion strategy that involves 'selective remembrance' focussed on intrinsic components of the former use such as architecturally distinguished buildings. To us, expressions of selective remembrance serve as a counterpoint to the silences of strategic forgetting and represent an important dimension of the agency exercised by stakeholders in the development process.

Re-imagining the Asylum as Housing Provision – Three Cases

We develop our arguments with reference to three exemplar cases – the former Sunnyside Hospital in Christchurch, New Zealand and the former psychiatric hospitals at Graylingwell and Knowle in Southern England. The drawing of exemplars from two countries allows us to gauge the interplay of national and community contexts, which was noted in Chapter 5 to be important in the re-cycling of former psychiatric asylums as tertiary education campuses in New Zealand and Canada. The choice of Sunnyside from within the set of nine former asylums in New Zealand was based on the fact that it is one of only three located within a major urban centre and the only one among these significantly redeveloped for housing (Kearns et al., 2012). We chose Graylingwell (near Chichester) and Knowle (located between Southampton and Portsmouth) because they are representative of the UK strategy of combining the preservation of heritage with the provision of affordable housing adopted by English Partnerships in 2005 (http://collections.europarchive.org/tna/20100911035042/ http://englishpartnerships.co.uk/hospitalsites.htm; see also *Daily Mail* 17/4/09). Sunnyside and Graylingwell share locations that were always more accurately described as suburban rather than rural, and now lie well within their respective built-up areas (Christchurch and Chichester), while Knowle remains representative of more rural (but not isolated) location. We look to the New Zealand case for insight into the process through which the stigma emphasised during the closure process is translated into attitudes toward the conservation of heritage and the memorialisation and remembrance (or strategic forgetting) of past use. From the two UK cases we seek to extract complementary data on remembrance and strategic forgetting during the redevelopment phase.

As noted in Chapter 2, our case studies feature a blend of observation and interpretation of textual discourse. Visits to each of the case study sites between 2010 and 2012 provided an opportunity to assess the current condition and use of sites and buildings and speak with developers. Access to discourse on the re-use of sites and buildings was obtained via analysis of media coverage, development reports and websites. Relevant articles for Sunnyside were extracted from *Newztext Plus*, an on-line database of full text copy contributed to by the publishers of New Zealand's major print media, supplemented by the use of microfiche archives to access more elusive documents. A similar approach using http://dailynewspaper. co.uk/, again supplemented by archival sources, was used to access relevant print media for Graylingwell and Knowle. An important characteristic of our research strategy is that it was comprehensive in scope. All available newspaper reports relating to re-development, rather than a sample of reports, were examined. In addition, we distilled from the reports the views of various stakeholders (developers, members of the local community and representatives of government) expressed at the time events were unfolding.

From Sunnyside to Linden Grove

The Sunnyside Asylum in Addington, Christchurch was opened in 1863 to house those considered to be insane and who had, until then, been held in the Lyttleton jail. Its main buildings were constructed in the 1870s from designs by Benjamin Mountfort, one of the country's leading Victorian Gothic architects. An equally impressive administration building, designed by John Campbell, was added in 1892. The New Zealand Historic Places Trust (NZHPT) noted the latter to be set in " ... an area of parkland, gardens and a fountain at its frontage" and to be ... "the public entrance for the Sunnyside Hospital from 1892 to 1999". The Trust also noted the cultural and architectural significance of the building: "[it] set an indelible memory of the institution in the minds of the thousands of patients, their family and friends who spent time in the complex. It is of outstanding significance as a unique remnant of a Victorian 'Lunatic Asylum' in the Gothic style" (Christchurch City Libraries, 2010). Starting with only 17 patients, the hospital had a peak population of more than 1,200 (Canterbury District Health Board, 2007). It officially closed in 1999. Figure 6.1 charts its subsequent redevelopment for housing.

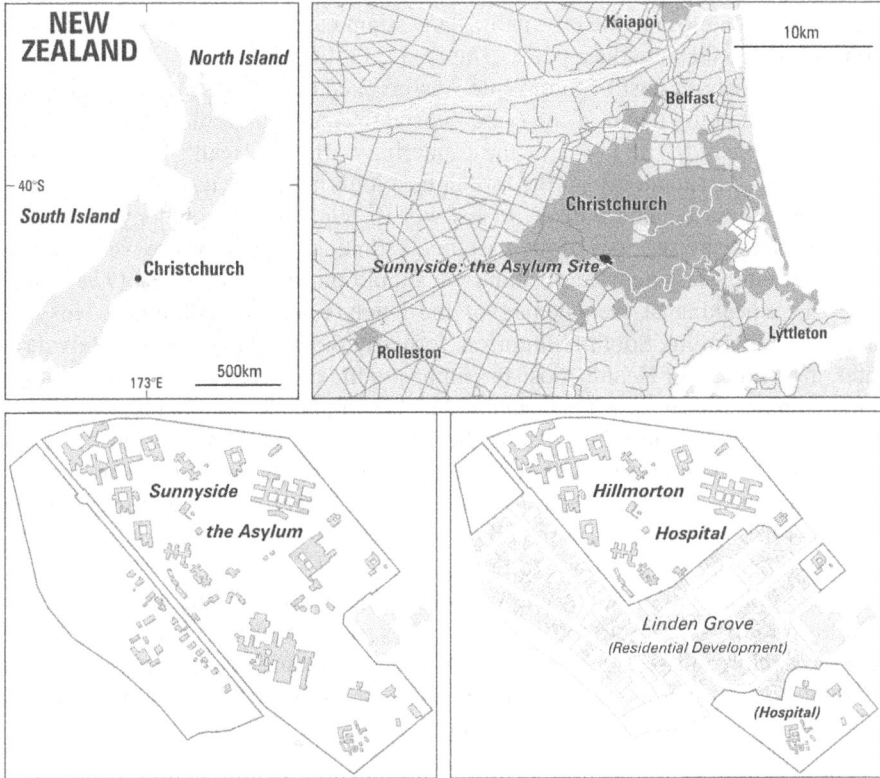

Figure 6.1   Residential Redevelopment at Sunnyside, Christchurch, New Zealand

The end of Sunnyside as a landmark within the health care landscape of Christchurch had, ironically, been preceded a year earlier by the announcement of a "multimillion-dollar redevelopment of Sunnyside Hospital" (*The Press*, 19/9/98). The mental health director of Healthlink South noted that "much of Sunnyside was constructed more than 60 years ago and the facilities reflect the historical ways health care was delivered". In the same article, it was reported that part of the site was surplus to needs and had been identified as such in the Ngāi Tahu Settlement Bill (a New Zealand Māori land rights determination), such that the Ngāi Tahu iwi (tribe) had first option to purchase it. A year later, the Sunnyside administration building was closed and the site was re-named Healthlink South Hillmorton. The mental health services managing clinical director noted that "[Sunnyside] was an asylum where you locked people up and threw away the key ... it stands for where psychiatry used to be, and not where it stands today" (*The Press*, 7/9/99). Six months later, it was proclaimed that "a 30-hectare area of land from the former Sunnyside Hospital site in Christchurch is to be sold complete with [low-density] residential zoning". A real estate source said the site was "good for housing, with a good location close to the city, and ... any stigma associated with the Sunnyside connection would soon disappear" (*The Press*, 13/4/00).

The stigmatised past of Sunnyside was invoked again, and more forcefully, a year later when the name of the (very substantial) hospital remaining on the site was re-considered. The Chair of the Canterbury District Health Board (CDHB) reported that he did not favour the former hospital's name because of the 'negative connotations' associated with it. In his words, "Sunnyside is synonymous with mental health care services of the past, and I think, no matter where we sit, it is a source of embarrassment, and I think we should put that behind us" (*The Press*, 12/3/01). Speaking at the opening of the newly renamed Hillmorton Hospital (formerly Healthlink South Hillmorton, formerly Sunnyside Hospital) a few days later, the Minister of Health chose to emphasise the connection between buildings and treatment: "Treatment for mental illness has graduated from being provided in these pre-1960 dark buildings that typified the bleak old asylum, to bright, modern, purpose-built facilities that will meet the needs of patients for years to come" (*The Press*, 16/3/01). In this statement we see graphic examples of the binary coding often evoked within deinstitutionalisation discourse such as light/dark and old/modern (Gleeson and Kearns, 2001).

More vilification of Sunnyside followed. Commenting on suggestions that the administration building should be preserved as part of a park project, the mayor of Christchurch wondered "why we would want to save something that is reflective of a lot of hurt in some people's lives" (*The Press*, 3/4/01). A few days later, a sharp comparison between the former and new hospitals was offered:

> The gothic institution looms ominously against the bright autumn sky. It's noon, but the old Sunnyside Hospital building seems shrouded in shadow. Many of its windows are smashed. Some have been boarded up and others have been left broken, allowing wind and rain to swirl unhindered into the gloomy interior ... Barely 500 metres away, the reason for the old grey building's abandonment is clear: a cluster of red-brick bungalows clump together in a cheerful huddle. The buildings are the new face of in-patient mental-health care in Canterbury ... There is no room for ghosts and spectres here (*The Press*, 7/4/01).

Thus, it was no surprise that three years later, and despite public input to the contrary, the CDHB confirmed that it would dispose of a 22.2 hectare site at the Hillmorton Hospital (*The Press*, 14/7/04), with Ngāi Tahu the likely purchaser. This decision shifted the focus of public debate to the fate of the architecturally-significant administration building that occupied part of the site offered for sale.

In April, 2006, in the absence of a private buyer, it was reported that "a historic Christchurch building will be demolished to make way for a Ngāi Tahu housing subdivision after the Historic Places Trust gave the green light to pull it down" (*The Press*, 5/4/07). After some months of prevarication on the part of the City Council, in November councillors "voted 11–3 against buying the building, saying that it was not prepared to commit to the $1m cost of buying it and renovating it to a useable standard" (*The Press*, 10/11/06). In the same article it was reported that half of the sale price set by Ngāi Tahu Development "reflected the value of the land and the negative impact of the 'imposing' building on the rest of the subdivision". It was noted that "the only chance for what heritage advocates say is the last New Zealand remnant of a 19th century asylum with Gothic features is if a private buyer is found in the next month" (*The Press*, 10/11/06).

The last days of the building were preceded by further vilification of its past, with the mayor of Christchurch exhorting "supporters of the last ditch bid to save the former Sunnyside Hospital administration building (to put) ... their energy into something worth saving" (*The Press*, 4/5/07), while a letter to the editor on the following day asserted that "for many mental health consumers, the darkly satanic buildings of the old Sunnyside were a thing to be avoided, even feared" (*The Press*, 5/5/07). On the same day, it was announced that "the former Sunnyside Hospital administration building in Christchurch was finally brought down to earth yesterday" (*The Press*, 5/5/07).

In the more than six years since demolition of the old administration building, a third of the former site of Sunnyside Hospital has been converted by Ngāi Tahu Property into the upper middle class suburban sub-division of Linden Grove (Bowring, 2010). The name is derived from the impressive stand of Linden trees that line both sides of one of the roads leading into the sub-division, a road that once led prospective patients and visitors to the Victorian Gothic core of the old hospital. Visitors unable to 'read' this remnant of the former asylum landscape would likely walk without remembrance through the small garden nearby, with its fountain and cupola, features that the educated eye would recognise as the remnants of the former gardens and administrative building of Sunnyside. On the internet, virtual visitors to Linden Grove would hear of the proximity of the development to the central city, of its array of established exotic trees, and generous park and reserve areas. What they would not learn is that the site was originally part of Sunnyside Hospital and that it is flanked on two sides by the successor mental health care institution, Hillmorton Hospital, a masking of both former and juxtaposed contemporary uses through strategic forgetting, with very limited scope for selective remembrance.

The most surprising detail in the strange chain of events that unfolded after that day in 1999 when Sunnyside Hospital was 'closed' and Hillmorton Hospital was simultaneously 'opened', was the decision to divide the Sunnyside site into three parcels and sell off the middle of the three. This middle parcel was the least encumbered with buildings, these having been demolished in 1979, 2002/3 (the Mountfort-designed buildings) and in 2007 (the Campbell-designed building), and incorporated a significant area of garden and open space. However, as Bowring (2010, p. 89) observes, "following the sale, the site was bisected by paling fences which defined the area for the Linden Grove subdivision being developed by Ngāi Tahu Property". The subdivision separates the extensive clinical complex of Hillmorton Hospital from its laundry and kitchens. Despite the idiosyncrasies described above, the successful (read 'profitable') re-development of part of the Sunnyside site as Linden Grove demonstrated the feasibility of the residential option.

The message was not lost on the owners of other former psychiatric asylum sites In New Zealand. As noted earlier, the former Ngawhatu Hospital (Nelson) is now the site of the Montebello Housing Estate. Additionally, the former Cherry Farm Hospital (near Dunedin) is being re-developed as Hawksbury, while proposals for residential development are coming into focus at Kingseat (south of Auckland) and, to some extent, at Seaview (Hokitika) (Kearns, et al., 2012). Additionally, Unitec, the tertiary education provider occupying the former site and buildings of the Carrington Hospital in Auckland (New Zealand) has, as noted earlier, proposed a redevelopment of a significant portion of the site for housing(see Chapter 5).

Creating Communities in Southern England

Our two English case studies, Knowle and Graylingwell, were respectively one of the two county asylums for Hampshire and the single county asylum for West Sussex. Graylingwell Hospital, the West Sussex County Asylum, served a largely rural county with an urbanised coastal strip from its location on the outskirts of the county town of Chichester. It was a relatively late creation among the county asylums, having been developed following the local government reorganisation in 1888 which saw the separation of East and West Sussex. Built in 1897 on acquired farmland, it formed, with a nearby Napoleonic-era army barracks complex, a block of public sector land use to the north of the town (English Heritage, 2006). Representative of a variant of the much-used 'compact arrow' asylum design of wards branching off a corridor (Cracknell, 2005), it also featured a stereotypically large water tower, a chapel, a theatre and separate residences for staff. In later years, newer buildings were added, scattered around its large estate. Figure 6.2 sets out its metamorphosis from an asylum to a housing estate.

Knowle Hospital, the 'First Hampshire County Lunatic Asylum', was built in 1852 on farmland purchased for the purpose by the County authorities. It was constructed in a relatively isolated location to the north of the then small town of Fareham, roughly equidistant between Portsmouth and Southampton, the two largest settlements in the area. Though originally intended to serve the entire County, its catchment was shrunk by the construction of other asylums in the county borough of Portsmouth (1879) and in the north of the county at Basingstoke (1912–1921). In its heyday it was therefore essentially a facility serving southern Hampshire and Southampton. It was nonetheless significant in scale. Its site covered some 40 hectares and it was built for 1000 clients; later overcrowding prompted the subsequent development at Basingstoke but in the late 1950s it still housed over 1500 people. It had its own rail station, extensive grounds and in its early days a pub, presumably for staff use (Anon., 1992). Figure 6.3 shows how it too changed from an asylum to a residential development.

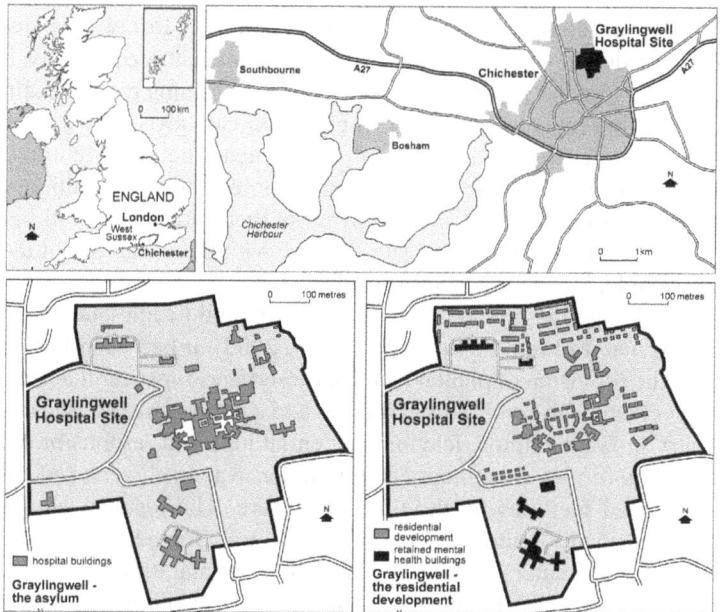

Figure 6.2   Charting change on the Graylingwell Hospital site, Chichester, UK

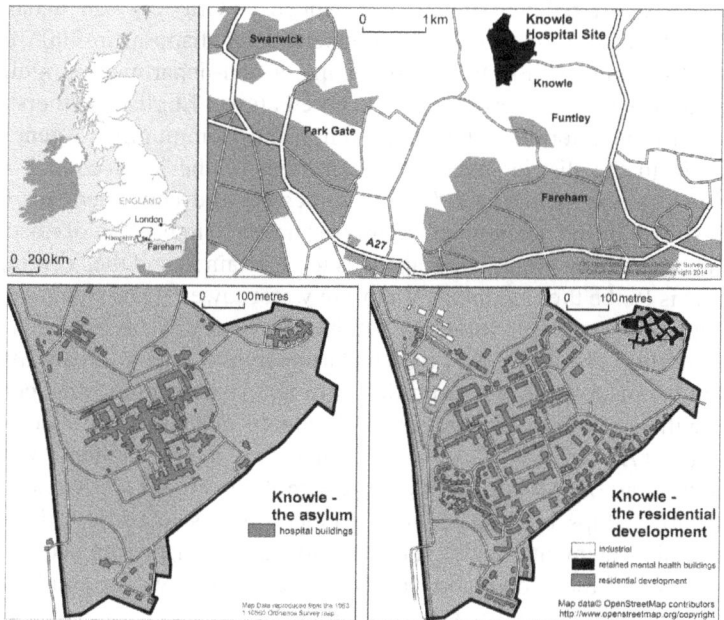

Figure 6.3   Charting change at Knowle Hospital, Fareham, UK

Graylingwell was closed in 2003 (Cracknell, 2005). Closure was marked by increasing dereliction as retained services were concentrated largely in newer buildings, mainly on the southern part of the site. Continuing mental health uses at the time of writing include the main mental health care provision for Chichester, child and low security forensic mental health near the south entrance and a separate mental health care facility for elderly people in an older building enclaved within the northern part of the site. In no sense therefore has mental health care disappeared from the Graylingwell site. Rather, with the exception of the facility for elderly people, it has moved to the edges.

A similar though less marked persistence of mental health land use is evident at Knowle. When it closed in 1996 (Cracknell, 2005), it had been in decline as a location for the provision of mental health care services for several decades. In the 1960s it lost clientele to community facilities and a new hospital-based department of psychiatry in Southampton, leaving residential inpatient responsibility for only a fraction of its former catchment. This function was itself replaced by smaller community-based facilities by the time of closure. The residual on-site mental health use at Knowle is now limited to the Central Southern England medium secure unit. This unit caters for clients with serious mental illness who need secure care because of their challenging behaviour or history as offenders. In comparison to Graylingwell, the residual mental health use at Knowle is thus perhaps more likely to sit uncomfortably alongside residential redevelopment though it too is located on the edge of the site.

The process of redevelopment commenced at Graylingwell in late 2006. The hospital site was one of the 96 transferred for disposal in 2005 from the former NHS Estates to English Partnerships, a non-departmental public body. Knowle had been an earlier disposal. As noted above, English Partnerships was itself later to metamorphose into the Homes and Communities Agency. Local guidance from the district planning department of the local council outlined a possible development of 1200 new homes on the combined hospital and barracks sites. For the hospital site there was an early indication of nostalgia for the former use. The asylum landscape was to be retained and the "centrepiece of the project is [to be the] landmark Victorian water tower" (*The News*, 16/11/06). It was felt that open spaces should be retained in what was envisaged as a mixed land-use setting. By 2007 the English Partnerships remit to create affordable high quality sustainable communities through regeneration was coming to the fore with proposals that Graylingwell should become a 'new eco-friendly community' (*The News*, 1/3/07). Agreement on redevelopment was finally signed at the start of 2008; initial plans were for 800 homes (*The News*, 30/1/08). English Partnerships brought together Linden Homes to address the building and conversion elements of the site plan, focussing on private sector housing provision, while the Downland Housing Association was commissioned to address issues of affordability and social mix.

Residential redevelopment began at Knowle in 1996, four years after its closure, and thus significantly predates that at Graylingwell. The main asylum building and several separate buildings were retained for conversion, largely in response to their heritage status. Other connecting buildings and the old south block were demolished. Over 150 apartments were developed in the main retained buildings and some 550 new houses were constructed on the former grounds. Approximately 20 per cent of the development was affordable/social housing. The lead developer for Knowle was Berkeley Homes, with the affordable and social homes element being a partnership with Thames Valley Housing. Berkeley Homes had extensive experience of the redevelopment of brownfield sites and board-level links with Crest Homes, a leading player in the redevelopment of several of larger former asylums in the London area.

Chichester District Council approved the proposed £250m redevelopment of Graylingwell in early 2009 after negotiation with local stakeholders and community groups (*The News*, 5/3/09). Permission was given for 750 'eco-homes', 40 per cent of which were to be 'affordable'. Provision was also made for community facilities, a school, shops and transport links. Construction started quickly and was scheduled over an eight year period (*The News*, 14/9/09). By early 2010 show-homes were open at what had become Graylingwell Park (Figure 6.4). Evocative of selective remembrance, commentaries sought to position the new housing in a historical (architectural) context and to stress the creation of a new community (e.g., "We first of all did a lengthy analysis of a typical Georgian house") and emphasised community connectedness (e.g., "We spoke to around 400 people in Chichester to find out what they wanted to see happen … . they all wanted somewhere for the community. So there are no fences, no barriers to the site") (*The News*, 26/3/10).

Similar objectives attended the redevelopment of Knowle as Knowle Village, although the earlier timing of its redevelopment meant it escaped the eco-town label and could more accurately be seen as a commuter development in which 'community' was linked with accessibility:

> [A]t Knowle Village in Wickham near Fareham in Hampshire, Berkeley Homes is building a range of new apartments and houses, with traditionally designed elevations with sharp contemporary interior layouts and fittings, together with a village green, grocery store, hairdressing salon and even a nursery school. This thriving new community, surrounded by lovely open Hampshire countryside, is just a mile from Wickham village and three miles from Fareham (*The News*, 28/11/07).

The redevelopments at both Knowle and Graylingwell both stirred controversy. Inevitably there were concerns at both sites about traffic and parking (*The News*, 28/5/09; *The News*, 16/6/10). At Knowle these concerns had a protectionist tinge in that they formed the substance of an objection by residents to the construction of a final phase of the development dominated by social housing. In this we can see how 'community' once created, becomes something to conserve, particularly if proposed changes threaten the existing status quo. Other controversies included the decision not to retain the asylum theatre at Graylingwell and the fate of the

Figure 6.4   Sales office, Graylingwell Park, Chichester, UK

chapel at Knowle, contentious until its donation by Berkeley Home for use as a community hall (*The News*, 27/5/04). The most significant controversies for our purposes in this paper were two that affected the Knowle site and a third that was specific to Graylingwell.

At Knowle both controversies were rooted in the location's repackaging as Knowle Village. In the local papers it was lauded in what, to judge by the repeated phraseology, were clearly developer-penned press releases (e.g., "[a] thriving new community, surrounded by lovely open Hampshire countryside, … just a mile from Wickham village and three miles from Fareham") (*The News*, 28/11/07). New residents were impressed: "Knowle Village is in a fantastic setting. There's a winding road leading to the village, which makes it feel as if we're escaping to the countryside, and our new home backs onto woodland so it feels very rural compared with our old home" (*The News*, 18/11/07). It was rural yet connected: "The M27 is just a few minutes' drive from Knowle Village offering easy access to Southampton, Portsmouth and the M3. Fareham train station, three miles away, provides a regular service to London Victoria in one hour 45 minutes". Contrasts were drawn against recent trends in urban expansion: "Modern day planners have often been accused of creating settlements which lack a community heart and are simply dormitory towns. But at Knowle, where hundreds of homes have been built, they have been striving to build around a village green style concept which will appeal to all age groups" (*Daily Echo*,

26/5/08). What we see in this prose is the regular deployment of signifiers of aspiration and accessibility alongside tropes of rurality, tradition and community and an underpinning of service provision.

In becoming a self-designated (new) village, Knowle thus became, for its equally new residents, a place that was expected to epitomise certain stereotypes of village life (Halfacree, 1995). Unsurprisingly, residents reacted negatively to proposals that challenged the idyll of " ... the first village to be built in Hampshire in more than a century" (*Daily Echo*, 26/5/08). The challenge came in the form of the Regional Spatial Strategy for the South East (GOSE (Government Office for the South East), 2009). This created a Fareham Strategic Development Area (SDA) that was intended to provide new housing in the area between Knowle and Fareham. It was a response to population pressures in the South East and threatened the vision of rurality and separation that the new residents at Knowle Village had embraced. Over 7000 new homes were anticipated and local councillors mobilised to fight the development: "It is important that we keep the green fields, especially around the Knowle section. We have always fought for Wickham and Knowle to remain countryside. We are only little villages but we will fight to make sure we do not become part of this big sprawl" (*The News*, 18/3/10). Although the change of UK government in May 2010 saw the abolition of the Regional Spatial Strategy, the threat to the village idyll remained alive as the local council and government more generally continued to seek ways to respond to population growth. In November 2010 the Village Residents Association indicated its opposition (http://www.knowle-village.co.uk/chair.html). Among its arguments were concerns about traffic, employment and waste disposal. They felt there was no local demand for more housing and lamented the "Loss of open green space, farmland and wild animal habitat". Planning inquiries in 2014 continue to raise the spectre of an end to the rural idyll.

A further issue raised in opposition to the Fareham SDA was the absence of " ... doctors, dentists, schools. All the things that were supposed to be in place for Knowle and we're still waiting for". This brings us to the second controversy affecting Knowle following its redevelopment as a 'village'. Not only was there concern about preserving the village. There was also concern about delivering the promise of the village. Like many new housing developments before, there was a gap between developer promises and provision on the ground. "When we moved here they said there would be facilities already in. We're paying council tax but there's not even a phone box or post box. They should at least have put up something like a [prefabricated] Portakabin shop" (*The News*, 8/4/02). In the words of another resident, "I feel we've been dumped in the middle of a field. We've been told we're getting boutiques but we want a convenience store and post office. And a pub isn't even on the horizon" (*The News*, 30/8/02). As one local councillor stated: "There are some pleasant aspects to the development but it has just got that unfinished feel" (*Daily Echo*, 26/5/08).

The vision of the village was also disturbed by disconcerting echoes of the past. As already noted, the regional secure unit is a continued presence on the site.

Figure 6.5   Knowle Village, Fareham, UK

Escapes and the threat of danger posed by the unit remain concerns for the new villagers: "Instead of medium security it should be maximum, especially with so many houses close by. My children love playing outside but I have been watching them all week since this [escape] happened" (*The News*, 20/8/03).

At Graylingwell thwarted visions and potential threats are issues for the future. At this location the controversy was over the name of the new development. Linden Homes proposed to rename the site 'Livingwell', paying homage to the 'well' in Graylingwell but alluding to English Partnerships' concerns with sustainability and the notion of the eco-town. The managing director of Linden Homes also stated: "We genuinely felt that given the site's former use as a psychiatric hospital, by naming the development Graylingwell it may have potentially negative connotations for local residents that might discourage potential purchasers" (*The News*, 5/1/10). This overt recognition of the potentially damaging commercial implications of the long shadow of the stigmatised asylum is not unusual. Many other former asylums have shed their past identities upon conversion to housing (e.g., Colney Hatch has become Princess Park Manor, and the Holloway Sanatorium became Victoria Park). At Graylingwell, local residents sought retention of the existing name arguing that "[it] is much older than the hospital was, going back hundreds of years and referring to the Grayling Well, still on the site, which is believed to have been used in Roman times as a water supply source" (*Chichester Observer*, 5/1/10). The local planning committee chairperson stated "Graylingwell was in existence many years before the hospital, but Livingwell doesn't actually mean anything". Linden Homes bowed to the popular view: "we have learned that [Graylingwell's] long history, which dates back to 1231 – considerably longer than its use as a hospital – carries tremendous weight with local residents and it quickly became clear that they would prefer we retain the site's original name" (*The News*, 5/1/10). We interpret these events as indicative of community and government stakeholder groups colluding in the deployment of selective remembrance by pointing to the distant origins of the name rather than its recent psychiatric deployment.

Such heritage concerns were a key theme in the entire development process at Graylingwell but far less evident at Knowle, despite concern over the fate of the chapel building. Frequent mentions were made of the 'Queen Anne style' of the main Graylingwell buildings and there was concern to ensure the preservation of the historic Grayling Well and, in an echo of the Knowle case, the asylum chapel (*The News*, 16/11/06). English Partnerships was clear that "We have been particularly careful ... to involve English Heritage to ensure the historical aspects of the site are respected" (*The News*, 5/3/09) and the joint development proposal claimed to be "Making the most of history: The team wanted to create a scheme that re-uses and enhances the historic buildings and the attractive parkland setting" (Anon, ND b). Heritage designations framed the possibilities for redevelopment, rather more than they did at Knowle. This was largely because the asylum parkland at Graylingwell was a designated Conservation Area and listed on the English Heritage Register of Historic Parks and Gardens. As English Heritage put it in their site assessment: "The provision of parkland to the west and at the South and North entrance to the grounds ... serves to reinforce this separateness from the outside world, as does the avenue of trees that formed the principal approach" (English Heritage, 2006). Knowle in its rather more rural and isolated setting did not function as a green lung for the nearby town. Indeed, landscape preservation at Knowle appears to have been a major issue in just two ways. First, that most traditional of bucolic signifiers, the cricket ground, was, in an echo of the situation at St James Portsmouth (Chapter 4) initially retained, though it was subsequently subject to development. Second and significant for our argument, the developers funded a secluded memorialisation of the asylum cemetery. In woodland adjacent to the new village one finds an interpretation board and a stone stating "Knowle Hospital Cemetery. Five and half thousand people were buried in this woodland between 1852 – 1971". This offsite memorialisation is the main acknowledgement of the former Knowle asylum.

Selling what was purported to be "the largest carbon neutral housing project in the UK" (Anon., 2008) was not, then, a case of strategic forgetting at Graylingwell. There was, however, rather more forgetting at Knowle. At Graylingwell a combination of heritage, community support, and the asylum landscape being seen as a resource, ensured instead selective remembrance. Knowle sold the rural but otherwise eschewed mention of the asylum; it manifested forgetting. In March 2008 a company of architects, community planners and urban designers organised a participatory planning weekend on the Graylingwell redevelopment. Their conclusions made clear a public view that "Key buildings should be retained and their history and heritage should be celebrated" (Anon, 2008). While 'celebrated' may be an example of marketing-speak it is also a word rarely associated with asylums. The reality of the marketing literature was more coy and had more in common with the Knowle experience: "Set in acres of breathtaking parkland and located less than a mile from the vibrant historic city of Chichester, Graylingwell Park is a landmark collection of beautiful homes and inspired character conversions" (Galliford Try Homes, ND). "Graylingwell Park offers you much more than just a home. With a welcoming community, sensitively landscaped parkland as far as the eye can see, artists' studios, a café, farmers' market and sports grounds for the children, you can look forward to an inspiring lifestyle for many years to come". Graylingwell asylum was gone but not, at least at this stage in its redevelopment, forgotten, unlike Knowle.

Reflections on Strategic Forgetting and Selective Remembrance

> At stake, are a series of the largest most remarkable and little-known public buildings in England built at great expense and set in superb landscape grounds which are often now in the full splendour of maturity. The quality of these buildings is the greater as so many were the subject of architectural competitions (SAVE 2005, p .1).

In reflecting on the foregoing narratives, we note considerable differences in the processes and outcomes associated with the UK and New Zealand cases. In the UK, discussions about re-use have both a national and a community context. Debates about the disposal of surplus public land and methods of coping with pressure for housing development are set alongside recollections of stigma. Heritage and planning legislation has also ensured the integration into new developments of significant aspects of the built form of the asylum as well as asylum estate landscapes. Strategic forgetting is evident in that potential buyers are often told that a site was formerly a hospital but not of its specialised role. There is also selective remembrance, albeit often implicit. Limited memorialisation occurs, quintessential asylum buildings and landscapes are preserved and names are, in some cases, retained. Graylingwell conforms almost perfectly to this general expectation; Knowle a little less so. In contrast, in the New Zealand case the stigma associated with asylum use was repeatedly invoked in what was allowed to remain a purely local debate. Indeed stigma threads its way inexorably through the narrative of closure and re-development at Sunnyside, either as the unquestioning valorisation of a move to community-based care (Gleeson and Kearns, 2001) that rendered Sunnyside and similar asylum facilities redundant, or, more starkly, as the active vilification of the institutional approach to mental health care and a consequent willingness to embrace a future that erased the past.

The bulldozers that removed the final traces of Sunnyside's emblematic built form in 2007 were removing stigma as much as a heritage building that had been allowed to deteriorate. A brownfield site wiped clean of large scale reminders of the past was clearly preferred by all the parties who worked to facilitate removal. Indeed, the developer Ngāi Tahu Property went as far as to put a combined price of $500,000 on the land occupied by the building *and* the stigma that would persist through its survival alongside the proposed housing development. Moreover, it is notable that the often-tenacious campaign to retain the former administration building was based on the architectural significance of the building rather than on the contribution of the institution to the life of Christchurch. We speculate that if the administration building had survived, it would have become the focus of strategic forgetting and selective remembrance in the same way that similar buildings, preserved selectively in the two UK case studies, metamorphosed into attractive 'period' apartments with little reference to their past. Indeed, to this end, it is intriguing to note the pressures at Graylingwell to preserve the water tower. From being an emblematic representation of the stigma of the asylum this edifice was repackaged to become a selling point for the new eco-village.

These divergent paths to redevelopment raise questions about attitudes toward the conservation of culturally or architecturally significant buildings in the two countries. New Zealand places considerable emphasis on the preservation and management of its natural heritage, and this is supported by far-reaching legislation and a well-developed system of national parks and other designated areas administered by the Department of Conservation (Bade, 2010). Such is this organisation's focus on biological heritage (arising perhaps from its roots in the former Wildlife Service) that it is often only the agency of volunteer activism that leads to the retention of examples of cultural heritage (Kearns and Collins, 2006). One might think that the previous loss of architecturally significant buildings at the Porirua and Seacliff asylums because of earthquakes and subsidence respectively would have resulted in national interest in preserving the administrative building at Sunnyside, but it did not. Indeed, the NZHPT played a strange role in the process – their 'accidental de-listing' of the building at the precise time that the development application was being assessed by the Christchurch City Council seems to be a most convenient coincidence that could, alternatively, be seen as an expression of agency. The demolition of Sunnyside's administrative building leaves the main building of the former Carrington Hospital as the only surviving example of Victorian asylum architecture in New Zealand (Kearns et al., 2012).

This valorisation of site but not buildings for residential re-development and the associated vilification of the former use in New Zealand are reminiscent of attitudes in the UK in the 1980s and 1990s, as reported by Lowin et al. (1998) and Franklin (2002). Our case studies suggest that this perspective no longer holds sway in the contemporary UK situation, and that this shift is underwritten by heritage legislation that protects both buildings and grounds and accepted, albeit sometimes reluctantly, by the development industry. Popular sentiment, while certainly strategic in its forgetting, is also selective in its remembrance.

The perhaps surprising value attached by the host Chichester community to the Graylingwell site and indeed to its very name (reaching back to pre-asylum times), was also linked to a remembrance of the position of the asylum as a familiar part of town life and a valorisation of the asylum's 'Queen Anne' architecture and its parkland. It may be that, as suggested by Franklin (2002), a sufficient passage of time is necessary for the opportunities that are inherent in combining residential development with heritage preservation to come into focus; such sufficient passage of time clearly did not occur in New Zealand where the process of asylum closure was both swift and tainted with controversy (Brunton, 2003).

Notwithstanding the differences in the re-development narratives noted above, especially in the invocation of stigma, we now see at all three case study sites an obscuring of the previous geography of the asylum. What remains is a rearrangement and fragmentation of its earlier form. At Sunnyside, the former asylum site is split into three parts and there is no adaptive re-use of the buildings. The quasi-normalisation of the asylum setting in the middle third of the former site is disrupted by ongoing mental health care uses to each side of it (see Figure 6.1). The Sunnyside name has given way to Hillmorton, but there are low-key instances of its memorialisation in the form of plaques, remnant gardens and cornerstones incorporating the former name in some of Hillmorton's older buildings. At Graylingwell, the retained mental health facilities that continue on the periphery of the redevelopment site have been renamed. They are *at* Graylingwell but no longer *of* Graylingwell; they are referred to by their road address or by new titles. The new Graylingwell Park is a repackaging of the past in which the former use is present but overshadowed by its future as an eco-community. Similarly the secure facility at Knowle is in Knowle Village but is known as Ravenswood. Like Hillmorton, Ravenswood is a renaming that symbolises a new era and set of norms (Berg and Kearns, 1996).

In the case of Sunnyside, the broader project of strategic forgetting began while the heritage administration building still stood, and was heralded by the name change – a mere but very effective stroke of the pen. This act removed what was effectively a marker of a previous era: a euphemistic name (Sunnyside) that acknowledged the pioneering psychiatric work carried out at Sunnyside Royal Hospital near Montrose, Scotland, and reflected the aspirational optimism of the past. Attaching the name Linden Grove at once provides distance from the former use and celebrates the Englishness of Christchurch (see Bowring, 2010). In contrast, Graylingwell Park exhibits elements of selective remembrance alongside strategic forgetting. It incorporates the name of the former asylum, but in marketing material the reader is encouraged to think of the well after which Graylingwell Farm was named. There is passing reference to the asylum in the decision to retain the name, but the (mental) hospital acknowledged to have existed for over a century and whose attractive core buildings are now incorporated into Graylingwell Park is presented as a fleeting episode in a longer, more naturalistic history. The retained buildings and grounds enable selective remembrance but their continued existence is folded within a dominant trope of selective forgetting.

Knowle also retained its name although otherwise its erasure of the asylum past was fairly complete. Indeed, in an interestingly geographical manifestation of strategic forgetting, it is now accessed by an entirely new entrance road; the former roads to the asylum now connect with each other, and with the secure unit, but do not enter the redeveloped site.

Conclusions

We can draw a number of conclusions in relation to our broader themes. First, the case studies invoke for us an unanticipated connection with earlier literature on deinstitutionalisation (e.g., see Dear and Taylor, 1982). In all three cases, mental health services remain on part of the original asylum site. Similarly, at all sites, but especially at the former Sunnyside where Linden Grove is sandwiched between the clinical and service zones of Hillmorton Hospital, largely middle class suburban populations have wilfully chosen to live in proximity to active sites of mental health care. A similar willingness to buy homes in proximity to ongoing mental health care delivery was observed at St. James Hospital Portsmouth (see Chapter 4). These populations are similar in demographic profile to those which vociferously opposed the location of the residential group homes and clinical facilities needed to support community-based mental health care (Dear and Taylor, 1982). This otherwise paradoxical juxtaposition can be explained on two counts. First, the housing at Linden Grove is new-build and arguably its contemporary style and considerable expense has transformed buyers into residents for whom the advantages of individual dwellings and the new neighbourhood outweigh any stigma by association and proximity. Second, the mitigations enacted by Ngāi Tahu Developments have clearly been more pronounced that those emplaced within the suburban communities of Dear and Taylor's Toronto. At Linden Grove, there are high paling fences that, in the words of Bowring (2010, p. 89) "impose a kind of violence on the site" and offer a "seemingly arbitrary bisecting of roads and gardens" (see Figure 6.6). Similarly, in Graylingwell Park new residents live in close proximity to active sites of mental health care but are separated from them by open space or hedges, while at Knowle separation is accentuated through the use of a separate entrance to the Ravenswood facility.

Our second general conclusion is with regard to memory, remembrance and memorialisation. While there is no explicit memorialisation on the site of the former Sunnyside Hospital, prompts for remembrance do exist. The most obvious of these prompts is the presence of Hillmorton Hospital, with its solid surrounding fence. In the park at the edge of the development, there are reminders of Sunnyside in the form of a restored fountain and a cupola incorporating a roof feature from the previous hospital, but these are signifiers of the asylum only to the informed. At Graylingwell, the park and the restored buildings are an omnipresent reminder but again perhaps only to those familiar with asylums. Much the same can be said of Knowle. At both UK sites mental health care facilities have been retained

Figure 6.6   The Separation of Linden Grove from Hillmorton Hospital on the former Sunnyside site, Christchurch, UK

and at Knowle there is clear separation as at Linden Grove. All three sites also feature chapel buildings which, though to varying degrees re-commissioned for other uses, continue to offer memories and hold an ability to strategically generate remembrance. We assume that this remembrance will fade over time unless shared, perhaps through the creation of what we think of as 'communities of memory' on the internet (see Chapter 7). What is more prevalent, perhaps, is the remembrance of local residents. One shared memories of employment as a nurse at Sunnyside when two of the authors encountered her walking a dog during a site visit in 2010. More broadly, members of the Chichester community clearly evidenced remembrance in their desire to retain the Graylingwell name. Such examples, while small-scale and fleeting, remind us that people will find tactics to evoke memory regardless of the enactment of strategic forgetting and/or selective remembrance (de Certeau, 1984).

Third, our case studies undoubtedly carry resonances with related research on the redevelopment of dockland, industrial and other brownfield sites (Chang and Huang, 2005; Edensor, 2005b; Green Balance, 2006; Summerby-Murray, 2002; Waitt and McGuirk, 1997). We see parallels with the limited body of work on the redevelopment of former military bases (Bagaeen, 2006; Clark, 2010; Tunbridge, 2004). Similarly important as sources of local employment, settings with heritage buildings and, in some cases, substantial preserved open spaces, former bases can also evoke feelings of 'otherness' through a secret past and possibilities of land contamination. In an empirical manifestation of this connection, Berkeley Homes, the redeveloper of Knowle, was also the leading party in the metamorphosis of HMS Vernon, a naval base in nearby Portsmouth, to Gunwharf Quays, a regional shopping destination and harbourside luxury housing development (Cook, 2004). While Gunwharf features missiles as street furniture, there are no complementary mental health signifiers at Knowle. A New Zealand parallel is the redevelopment of the Hobsonville Airbase in Auckland into Hobsonville Point, which is being marketed as an 'affordable coastal community', complete with a 'Catalina Cafe' and other heritage-coded facilities and street names (Opit and Kearns, 2013). In these cases there is arguably an ease with which military (in contrast to mental health care) heritage is 'played up' in the re-development process and incorporated into broader aspirations of urban distinction. There are also resonances with research on gated communities (Atkinson and Flint, 2004; Grant and Mittelsteadt, 2004; Le Goix and Webster, 2008), echoing our earlier observation in Chapter 3 that in the re-imagining of psychiatric hospital spaces the therapeutic values of asylum and isolation can potentially be exploited in the quest to keep "a troubling world at bay" (Joseph et al., 2009, p. 86). Former Victorian lunatic asylums are often seen as favoured sites (Weiner 2004; Manzi and Smith-Bowers, 2005), with Blandy (2006, p. 21) regarding them as desirable because they " ... often include a large perimeter security wall ... satisfying the desires of those wishing to live in gated communities".

In closing, we return to the tension between heritage conservation and strategic forgetting and selective remembrance within the re-development of former psychiatric asylum spaces. Through attention to the context and detail of three case studies, we have offered a close consideration of the challenges inherent in the transformation of former psychiatric asylum sites to housing use. We have shown that the long shadow of past use can be acknowledged both through its formal memorialisation and through (selective) remembrance triggered by the survival of aspects of distinctive vernacular architecture and site plans. In particular, we have noted a varying embrace of the asylum past.

While we have focussed on the particularities of three case studies (drawn from the UK and New Zealand), psychiatric asylums are commonly found on the edges of many Western cities and, as they have closed, some have offered rare opportunities to paradoxically reorient their use from offering secluded, separate and contained care for patients to offering the same characteristics of seclusion, separation and containment from the rest of the city for affluent house buyers. Given this contradiction, we see these sites as liminal spaces – not only in the sense of being at the edge of the city, but also as 'edgy' spaces to the extent that the shadow of their former use must either be embraced, transformed or suppressed. Suppression, which we have termed strategic forgetting, is the most common approach to the re-imagining of former asylum spaces as residential developments. While helping to underwrite the preservation of listed buildings and valued landscapes, the combination of heritage conservation and housing development does not guarantee the remembrance of the asylum within the cultural landscape because of the strong tendency toward selective remembrance. In this sense, we wonder whether future generations will appreciate the preservation of heritage architecture and landscapes without remembering the asylum use for which they were both created. Reaching back to the ideas of Nora (1989) introduced in Chapter 2, we wonder whether the memory of the asylum, regardless of even the broad-scale preservation of buildings and landscapes, will survive the passing of those who, as patients or staff, populated these psychiatric spaces. It surely will, but without much of the nuance and local flavour that we currently encounter. The link between memorialisation and remembrance will weaken, as the locus for the latter shifts from the recycled and transformed sites of former treatment to alternative spaces such as the internet. This notion of virtual remembrance forms part of the narrative surrounding derelict asylums and it is to that fate that we now turn.

Chapter 7

Imagined Geographies and Virtual Remembrance at the Derelict Psychiatric Asylum

This chapter considers the fourth of the fates posited for former psychiatric asylums in Chapter 1 – dereliction. We see this fate as literal and metaphorical, ephemeral and legacy-making. In a literal sense, dereliction involves the polar opposite of retention of health care use on site, our first fate (see Chapter 4). Abandonment renders sites and especially buildings prone to the ravages of weather, vandalism, neglect and infestation by animals and plants such that dereliction becomes increasingly evident. However, dereliction can be ephemeral, and may be terminated by three intertwined processes: capitalisation of value (through redevelopment); demolition (leading to a future use as open space or as a prelude to new development); and restoration (investment in the asylum landscape or buildings as heritage). In a metaphorical sense, in an abandoned state these derelict sites and buildings serve as spectral reminders that evoke remembrance of what was once mainstream and is now 'other'. In contrast to the guided forms of remembrance through overt or discreet forms of memorialisation that may occur on sites re-purposed for other institutional uses (Chapter 5) or residential purposes (Chapter 6), dereliction gives free rein to the imagination of the observer. Sites and buildings will also often be evocative of the forbidding and haunting imagery of the gothic asylum in popular culture. Indeed, through this alignment with the imagined rather than the real, the memories evoked by the derelict asylum may become paradoxically appealing in the remembrance of this now outmoded form and site of treatment in the sense that they 'send a shiver up the spine'.

Below, we focus on ways of seeing the derelict psychiatric asylum and its stigmatised past. We ask how former asylum use is remembered in the specific context of derelict asylum sites. This is a timely research question. Dereliction is encountered in each of the countries with which we deal extensively in this book. For example, in our stocktake in New Zealand, we noted two of the country's nine former psychiatric asylums to be in a state of complete dereliction, while three more were found to be in a state of partial dereliction (Kearns et al., 2012). Yet dereliction has been, at best, a peripheral theme in our earlier chapters, despite a theoretical resonance that might see asylums as former spaces of psychiatric care that are haunted by their past. The presence of derelict asylums in the contemporary cultural landscape is a spectral reminder that evokes memories and images of a stereotyped and often demonised history. In dereliction, we argue, asylums have become places where security fences and patrols enhance a sense of 'otherness' that can both repel and attract.

The remainder of the chapter is organised in four major sections, the first of which elaborates upon the discussion of theoretical framing and analytical approach set out in Chapter 2. We then shift our focus to the substantive research material. Specifically, we draw on the commentaries of individuals who explore abandoned asylum buildings. Visual imagery, blogs and internet forum discussions in the UK, Canada, USA and New Zealand are examined in order to reveal complex and nuanced understandings of the derelict asylum. These materials speak to the identification of places that, while simultaneously sites of play, danger and discovery, are also locations where particular images of the psychiatric asylum are recovered. The chapter concludes with a discussion of the relevance of these virtual narratives for the framing of collective memories of the psychiatric asylum.

Framing Research on the Derelict Asylum

Given the stigmatised past of the psychiatric asylum, one tenable approach to interpreting these perceptual understandings and imaginings would be to draw on the ideas of strategic forgetting and selective remembrance outlined in Chapter 2 and applied extensively in Chapters 5 and 6. These complementary concepts illuminate the creative tension between the retention of positive memories of the asylum and the simultaneous obscuring of more negative aspects of that past. They are notions that are grounded in ideas about memory, remembrance and memorialisation (Halbwachs, 1992; Nora, 1989). Work on these themes related to more general issues of dereliction, especially in urban landscapes, helps to widen this theoretical lens. From within geography, research by Edensor (2005a) is particularly relevant, being concerned with ideas that help to reinsert (past) life into locations that are (presently) derelict. His examination of relict industrial sites in northern England sees the dereliction " ... conjur[ing] up the forgotten ghosts of those who were consigned to the past upon the closure of the factory" (2005c, p. 311). This is a close analogy of our concern with the stigmatising past of now derelict asylum spaces.

We also draw inspiration from work on psychogeography and spectral geographies. In its contemporary form psychogeography combines subjective and objective knowledge to develop creative understandings of places. Despite its name, it has attracted limited attention within the discipline of geography. Bonnett (2009) provides one exception, focussing on the writings of psychogeographer Iain Sinclair. Bonnett does not dwell on Sinclair's fascination with former asylum sites (Sinclair, 2002) or indeed with madness in general (Sinclair, 2006) but nonetheless he effectively identifies the redemptive qualities Sinclair sees in remnant spaces and histories as indicative of "cultural and social loss" (2009, p. 45). He also notes how remnant spaces are potentially unsettling, having what we might term spectral qualities in the sense that there are discernible presences of the past that are evident in the present. Work on such spectral geographies, examining the shadows cast by past events, has been rather more popular within geography than psychogeographic research with some referring to a 'spectral turn' (Holloway and Kneale, 2008; Luckhurst, 2002; Maddern and Adey, 2008). Wylie (2007), in particular, has signalled a Derridian concern for how places are unsettled by haunting, while Till (2005) speaks of the interplay of haunting and nostalgia. For our part we anticipate that consideration of derelict asylums will unearth themes of remnant presences and shadows of past practices at what can be unsettling sites of abandonment and dereliction.

Finally, we look for inspiration to literature on the theme of dereliction. There is a considerable literature on derelict land, housing and 'brownfields' both in geography and in planning studies (e.g. Dear, 1976; DeSilvey and Edensor, 2012) and a small but significant literature on derelict industrial buildings, their redevelopment and their role in problematising notions of heritage (e.g. Atkinson et al., 2002; Hamnett and Whitelegg, 2007). However, our interest focuses on the emergent body of work on individual encounters with dereliction charting the performance and practice of urban exploration (Bennett, 2011; Garrett, 2010; Garrett, 2011; Garrett, 2014; Pinder, 2005). Sites of dereliction, including former asylums, offer particular opportunities to urban explorers, individuals who perform contemporary acts of exploration in (generally) urban settings, indulging in "a cultural practice of exploring derelict, closed and normally inaccessible built environments" (Garrett, 2010, p. 1448). Whereas 'actively' inhabited buildings are predictable in what they offer, derelict sites provide intrigue and rawness, signs of past occupation, and challenges of entry. Such attributes make derelict sites and buildings such as asylums, a prime target for those with a sense of adventure and willingness to risk physical expulsion or even prosecution.

While the unstructured character of urban exploration has been commented upon (Bennett, 2011), it is generally held to involve engagement with "material and immaterial, functional and fantastical, rational and irrational histories of places" (Garrett, 2011, p. 1053). In this sense explorers create myths about places that are embedded both in their own and others' understandings of those places. For us, the performance of urban exploration bring life to the otherwise closed and redundant spaces of the asylum. Further, urban explorers frequently extend the life of such sites through on-line galleries that may remain visible after advanced dereliction, demolition or re-purposing. Urban exploration offers a vivid account of the internal experience of dereliction. The key question that follows is how exploration in asylum settings is shaped and framed by the asylum past while at the same time shaping the remembrance of that past.

An Analytical Approach: Exploring the Derelict Asylum

Though we would not construct them as urban explorations *per se*, our own visits to the former Kingseat Psychiatric Hospital, approximately 45 km southwest of Auckland, New Zealand were the immediate prompt for our interest in dereliction and subsequently served to sculpt our approach to examining other cases of derelict asylums. Kingseat now comprises a mix of dereliction, ephemeral housing in decaying buildings and, since 2005, a very specific appropriation of the popular demonology of the psychiatric asylum through the re-use of one building as 'Spookers', a horror-themed visitor attraction (Joseph et al., 2009) (Figure 7.1). 'Spookers' has rescued Kingseat from a 'strategic forgetting' in the public consciousness and its naming capitalises on the foreboding character of the disused buildings (Kearns et al., 2012). Through the requirement that patrons drive a circuitous route through the asylum grounds to reach its entrance, Spookers harnesses a potent mix of repulsion and intrigue into a packaged opportunity to be frightened by choice.

Figure 7.1   Spookers Signage, former Kingseat Asylum, Auckland, New Zealand

Spookers also plays into a wider discourse of abandoned asylums as haunted places (e.g. Williams, 2008). Although it avoids trading overtly on its asylum past, the journey through the former hospital grounds nonetheless heightens the anticipation of horror. The visitor experience is further escalated by the frisson of entering a former asylum building. Further, on the Spookers website (www.spookers.co.nz), its haunted house attraction page features rendered images of shadows passing over a distant mysterious building on a hill: a stylised composition suggestive of an approaching encounter with somewhere Other. Thus customers enter first the grounds and then the building with preconceived expectations of ghoulishness and the macabre. Being on the otherwise derelict site of a former asylum is central to the experience. As Light (2009) has argued with respect to Transylvania and the Dracula myth, visitors do not simply encounter Transylvania; rather they *perform* a stereotype of Transylvania. We contend that visitors to Spookers similarly perform and recycle a stereotypical asylum experience prompted by locational associations and surrounding dereliction.

Our experiences at Spookers/Kingseat pointed us to three methods for researching the derelict asylum. First, it was clear that we needed accounts from people who were accessing abandoned asylum spaces and had direct familiarity with our target subject matter in its derelict state. These accounts would provide insights into encounters with dereliction that could be examined for understandings and stereotypes of the asylum past. Second, we sought visual images of dereliction. The visual iconography of the asylum in terms of its buildings and landscapes has been widely noted (Brown, 1980; Payne and Sacks, 2009). The dereliction of such abandoned mental health care settings might be expected to attract a visual record while simultaneously providing evidence of enduring themes from the asylum past. Third, our Spookers/Kingseat experience indicated the need for personal field visits to derelict asylums, both for observational purposes and also to substantiate their current state.

We chose internet accounts of urban explorations of asylums as our key source with respect to the first two methodological desiderata. These accounts convey to the reader highly personalised contemporary narratives of sites in a state of abandonment; they are a record of the asylum in its 'found state' of growing dereliction. The accounts mingle text and photography. Sometimes the textual element is brief and colloquial. On other occasions, blogs and forums develop extended narratives covering such themes as the challenges of entry, the rationale for exploration and the thrill of discovery. Photography similarly ranges from shaky snapshots to exhibition-quality imagery and from records of 'found-scenes' to tableaux in which explorers engage in posed encounters with their location. Internet host sites range in scale from those of national and international networks of urban explorers reporting on multiple locations to those focussed on single asylums or the 'journeys' of individual explorers. To extend our coverage, we supplemented material from urban explorer sites with asylum-focussed material from sites devoted to paranormal activity and ghost hunting. These latter sites focus overtly on the popular construction of the asylum as a haunted place and, to this end, we acknowledge a need to interpret with care accounts and imagery that range from the credulous to the sceptical and from the comedic to the pseudo-scientific.

The potential corpus of research material was vast. To manage this part of our investigation of the former psychiatric asylum, we focussed on material concerning sites where we had an existing familiarity. This decision ensured that we were aware of the development history of the chosen sites, the stories of closure and any background controversies during their periods as asylums. We supplemented the material on these known sites with equivalent materials on two 'type sites': asylums that have secured a high profile in urban exploration and have concomitantly high levels of documentation. Table 7.1 summarises the chosen sites. Collectively, they enabled us to illustrate the key themes evident in envisioning the derelict asylum.

Table 7.1   Selected derelict asylum sites

Asylum	Country, City	Maximum Occupancy	Closure Date
Graylingwell	UK, Chichester	>1,000	2001[1]
Cane Hill ('type site')	UK, Croydon	c. 2,000	1991[2]
Lakeshore/Mimico	Canada, Toronto	1,391	1979
Kingseat	NZ, Auckland	c. 800	1999
Carrington/Oakley	NZ, Auckland	494	1992[3]
Danvers State ('type site')	USA, Massachusetts	>2,000	1992

Notes: [1] Retained mental health services continue on part of the site. [2] An on-site secure unit remained open until 2008. [3] An on-site secure unit continues on part of the site.

As outlined in Chapter 2, we followed standard qualitative methodology in working with our textual and visual research materials (Silverman, 2010). We viewed our materials as 'text' setting out the views that urban explorers have of derelict asylums. Our objective was to subject this text to a critical reading. In this sense we were building on work that has used photographs as a tool for understanding places and memories (Goatcher and Brunsden, 2011; Hoelscher, 2008), methodologies for extracting emotional and affectual responses to the past (Pile, 2009), and approaches to understanding how landscape, performance and affect combine to offer ways of thinking about positionality Throughout our analysis of data we gave attention to interconnections between the written and the visual. From this process we identified key tropes within the materials that constitute common themes that could be placed within the context of wider literatures and theoretical positions introduced earlier in the book. In recognition of the unclear and variable copyright situation regarding the re-use of imagery available on the internet, we have limited ourselves to referencing the URLs of illustrative material in the discussion that follows and to the use of our own photography. Quotes from internet sites are attributed but presented verbatim; all sites were checked at the time of writing.

Our field visits enabled us to update knowledge and enhance our understanding of the sites in the light of our readings of the accounts of urban explorers. All visits were conducted in collaborative combinations and entailed photography, mutual discussion and subsequent note-taking, and interviews with individuals encountered on-site. Though there were some parallels, in that we were stopped on occasion by security personnel, our visits could not be described as explorations in the sense conveyed in the urban exploration literature. We sought out and investigated sites purely for research purposes and were open about our presence to the extent that we sought out other people on-site for conversation and discussion; to this end these visits represented an affirmation of the utility of observational fieldwork (Kearns, 2000). Nonetheless, in leading us to step onto abandoned sites, our visits gave us an empathetic understanding of the nature of urban exploration in the asylum context.

Encounters with the Derelict Asylum

We now turn to an analysis and discussion of our research material. We identified four key tropes that characterise encounters with the derelict psychiatric asylum. We examine each in turn:

Iconic Constructions

Though it may be a truism that asylum landscapes are instantly recognisable to the knowing enthusiast (Joseph et al., 2013), it is undoubtedly the case that explorers' encounters with asylums are characterised by tropes that might be described as iconic, deeply linked to the stereotypical vision of the asylum and signifying its presence (Table 7.2).

Table 7.2   Iconic constructions: sample research material

Theme	Site	URL
Water Towers	Graylingwell	http://www.derelictplaces.co.uk/main/showthread.php?t=16599, Image 21
	Cane Hill	http://www.28dayslater.co.uk/forums/asylums-hospitals/31752-cane-hill-tower-last-24–06–08-a.html, Images 1–6 http://www.simoncornwell.com/urbex/hosp/ch/e0808–2/32.htm
Frontal elevations	Danvers State	http://www.uer.ca/locations/show.asp?locid=20017
Driveways	Kingseat	http://meansxnomeans.tumblr.com/post/28617974905/abandoned-kingseat-asylum-auckland-new
Roofscapes	Cane Hill	http://www.forbidden-places.net/urban-exploration-cane-hill-asylum#gal
	Graylingwell	http://www.28dayslater.co.uk/forums/asylums-hospitals/54503-graylingwell-asylum-water-tower-october-2010-a.html, Image 10 http://www.derelictplaces.co.uk/main/showthread.php?t=3294
Dereliction & incursion of nature	Cane Hill Lakeshore Graylingwell	http://www.abandoned-britain.com/PP/canehill/12.html http://www.simoncornwell.com/urbex/projects/ch/ue/cam1.htm http://www.uer.ca/locations/viewgal.asp?locid=24935&galid=19991 http://www.talkurbex.com/forum/viewtopic.php?f=8&t=3339 (Image 4)
Stairwells	Graylingwell	http://www.28dayslater.co.uk/forums/asylums-hospitals/68304-graylingwell-asylum-chichester-jan-2012-a.html?
Corridors	Cane Hill	http://www.contaminationzone.com/Gallery19A.php
Looking out	Graylingwell	http://www.derelictplaces.co.uk/main/showthread.php?t=18481, (Images 2 and 3)
Tunnels	Cane Hill Graylingwell Danvers State	http://www.28dayslater.co.uk/forums/asylums-hospitals/37071-cane-hill-asylum-snow-february-3rd-2009-a.html http://www.28dayslater.co.uk/forums/asylums-hospitals/56727-graylingwell-asylum-hospital-chichester-west-sussex-january-2011-bit-wide.html (Image 13) http://urban-exploration.wonderhowto.com/inspiration/danvers-state-insane-asylum-massachusetts-luxury-living-0117698/

Prime amongst these iconic signifiers is the water tower in the UK, most famously seized upon as a symbol of the asylum by Enoch Powell in his eponymous 'water tower' speech announcing the end of the asylum era (Powell, 1961):

> There they stand, isolated, majestic, imperious, brooded over by the gigantic water-tower and chimney combined, rising unmistakable and daunting out of the countryside – the asylums which our forefathers built with such immense solidity – to express the notions of their day.

Figure 7.2   Graylingwell water tower, Chichester, UK

For urban explorers, the water tower is thus understandably a key feature within both written and photographic accounts. It is a subject to be photographed as a confirmation of the explorer's presence in an asylum, whether in the desolation and chaos of an asylum in the process of partial demolition, as at Graylingwell (Figure 7.2), or in the chaos of the encroaching dereliction of Cane Hill in southern England. The sight of the water tower signifies arrival in another land. In the words of one explorer of the latter site:

> I was pleased to see the water tower on the horizon as we drove down the A23. After stopping on the Portnalls Road, we found an overgrown public footpath. We walked down the wide muddy bridleway, screened by high bushes and overgrown grasses on either side. We turned the corner and there it was. A ward of the derelict, ex-lunatic asylum loomed menacingly over us. (http://www. simoncornwell.com/urbex/projects/ch/intro/index.htm)

Much the same interpretation can be attached to a second classic architectural feature: the frontal elevation of the main asylum buildings. Whether viewed from afar or from some other position, the resultant imagery conveys clearly the scale and monumental nature of the asylum: in Powell's words, "majestic ... imperious ... unmistakable" but above all "daunting". Nowhere is this clearer than in photography of the now demolished Danvers State Asylum in Massachusetts where the Gothic design of the gigantic frontage reinforced the forbidding aura of the former hospital (Figure 7.3)

For the intrepid explorer, however, the goal is to penetrate the interior of the asylum. As with traditional explorers there are routes to be pioneered, markers to be followed and destinations to be reached. Here we can again discern asylum iconography at work. If the water tower is a clear manifestation of the external iconography of the asylum, ascending the tower is an Everest of asylum exploration. In the words of one explorer:

> With the water tower under the watchful eye of security, stealth and cunning was required to get across 'no mans land' to the base of the tower. We spotted a chink in the towers armour and choosing the right moment we were inside. Once up top, it gave us a cracking overview of the hospital and some pretty good views to boot ... . We came, we saw, we conquered. (http://www.28dayslater.co.uk/forums/ asylums-hospitals/54503-graylingwell-asylum-water-tower-october-2010-a.html)

Once ascended, the photographic opportunities are legion, conveying images both of the world beyond the asylum, the outside world-beyond-the-walls, and also the complex, often confusing roofscapes of the asylum. Back within the asylum, the explorer must navigate *terra incognita*. Confusion and complexity are again evident. Photographs capture stairs and corridors leading to unseen destinations and in these we discern metaphors for patient careers, whereby they ascend stairs from shadow to light. These effects are heightened when accompanied by encroaching

Figure 7.3   Danvers State Hospital demolition, Danvers, USA

vegetation, and other evidence of decay. Occasionally, the photographic journey through the asylum halts at a vista of the world beyond captured through the windows of wards or in the views from 'airing grounds' where patients might once have contemplated life outside the asylum. On other occasions the journey continues to a further iconic presence: the vertical opposite of the water tower, the asylum underworld. Descriptive accounts of asylums are replete with references to tunnel systems. The narratives and photography of urban explorers mirror this fascination. Simon Cornwell's website recounts a journey through the tunnels beneath Cane Hill:

> Just past another manhole cover, the style of the tunnel changed again. It retained the height but with the two pipes now lay side by side, then walking down this tunnel was achievable only in a crab-like, sideways shuffle. We made the assumption that this tunnel just serviced the remaining wards on this side of the hospital and would get smaller. The more interesting parts of the network, and the tunnels to the boiler house, would've been back by the kitchens. So, we pushed open the manhole and made our way back into the corridors of Cane Hill. (http://www.simoncornwell.com/urbex/hosp/ch/e130702/116.htm)

At the risk of over-interpretation we discern in this text the revelation of hidden worlds within the already obscured world of the asylum. Elsewhere photography records the subterranean depths of the asylum showing receding perspectives stretching into the darkening distance. Our own encounters with the tunnel systems in Lakeshore (Figure 7.4) and Carrington were similarly atmospheric, revealing hidden storage areas holding material reminders of the former asylum use notwithstanding the converted and refurbished tertiary education buildings that stand above (Kearns et al., 2010)

Figure 7.4   Tunnel system, former Lakeshore Hospital, Ontario, Canada

Psychiatric Signifiers

Towers and tunnels may be desired destinations for the explorer of the derelict asylum. They may even be iconic signifiers of the asylum, but they are not the El Dorado of asylum exploration. For this we turn to psychiatric signifiers; locations that need little or no interpretation when it comes to identifying a link to mental health care (Table 7.3). This trope within our evidence base is represented by electro convulsive therapy (ECT) or lobotomy suites, pharmacies, morgues, the presence of hospital equipment and, *par excellence*, the padded cell. In engaging with these features, the photographic and written record touches directly on relics of the asylum past, on the performance of psychiatry within the asylum rather than architectural iconography, an envisioning largely cast in negative terms but also on occasion capturing more positive remembrances:

Disheveled corridors here echo and some of the claustrophobic isolation cells can only hint at some of the former treatments once administered behind hidden walls including electro-shock therapy treatment, various forms of early hydrotherapy and with more progressive treatments practiced here such as light and art therapy. (http://www.contaminationzone.com/Gallery19A.php (paragraph four))

Table 7.3   Psychiatric signifiers: sample research material

Theme	Site	URL
ECT suite	Graylingwell Cane Hill	http://www.talkurbex.com/forum/ viewtopic.php?f=8&t=3339 (Image 3) http://www.simoncornwell.com/urbex/ hosp/ch/e130702/96.htm
Lobotomy suites		See text
Pharmacies	Cane Hill	http://www.forlornbritain.co.uk/canehill.php
Morgues		See text
Hospital equipment	Cane Hill	http://www.forlornbritain.co.uk/canehill.php
Cemeteries	Lakeshore Danvers State	http://www.asylumbythelake.com/ cemetery/project.html http://www.uer.ca/locations/viewgal. asp?picid=182636

Within our chosen asylums, there are no longer examples of padded cells or obvious sites of past lobotomies. Danvers State is however celebrated for its links to the development of lobotomisation. Although the material remains are no longer extant, memories are easily evoked in the explorer community:

> … very sad place and the vibe that the building gives off is almost satanic i think … my good friend used to work at danvers state until the day it closed. he saw the horror that went on in that place he was the one who first told me how to navigate the building becuase of its imense size. he told me about the doctors labotomizing people just because of over population. by the way if you didnt know frontal lobe lobotomy was developed at danvers something i had thought about many times while walking through. (http://www.opacity.us/image3354_ backside_panorama.htm)

The febrile imagination casts asylum morgues as the haunting endpoint in a story of failed barbaric treatments and patient suicides that has filmic echoes in *One Flew over the Cuckoo's Nest* or *Shutter Island*. Indeed they serve as prompts for wilder imaginings:

> I know that The Nurses hostel in Kingseat has been transformed into Spookers haunted attractions, but I did some research on the main part,

where the ward is, and saw that 72 people had died there. With such a history, I wonder if it is haunted. I wouldn't mind renting one of the main parts for a while with some friends and seeing if I can catch anything. I also wouldn't mind checking out the morgue hidden out the back down in the bush. (http://nzghosts.freeforums.org/kingseat-psychiatric-hospital-t624.html)

Morgues offer uncomfortable reminders of the links between eugenics and mental health treatment (Phelan, 2005). Prosaically, but realistically, they prompt, alongside the asylum cemetery, a recognition of the inevitability of death within the often very large populations of asylum institutions.

There are other more mundane psychiatric signifiers capturing the medicalisation of mental illness and linking the realities of changing treatment technologies to popular imaginaries of the asylum. Pharmacies, for example, in various states of abandonment, serve as a reminder of the psychotropic formularies that were a necessary part not only of the therapeutic treatment of the asylum residents but also of the maintenance of control:

Amazingly the original dispensing cupboards were still in place along with this long table. I guess this was where individual patient's prescriptions were created. (http://www.simoncornwell.com/urbex/hosp/ch/e130702/38.htm).

We note a paradox in this element of the exploration narrative. Though drug therapies were central to asylum life, particularly as asylums succumbed to overcrowding, they ultimately enabled the development of the community-based mental health services that replaced the asylum (Gronfein, 1985).

Images of control and the technologies of health care are particularly evident in the fascination that urban explorers vest in electro-convulsive therapy (ECT). The ECT suite arguably ranks below only the padded cell and the lobotomy table as a destination of choice within the exploration trope of psychiatric signifiers. ECT offers a modernist technology of control and a potent combination of equipment, power and images of experimentation. Signage is often photographed as a precursor to discovery. Thereafter, the focus shifts to electrical equipment. Such is the single-mindedness of exploration that anticipation can overtake reality. If what is frequently photographed and discussed only resembles ECT equipment it may be of little matter; the exploration is as much an imaginary journey as a real encounter with genuine artefacts:

We found another ECT unit. A much smaller piece of kit, this was the handy, portable, battery driven "Transindolor B225", with attachments for two sets of electrodes.

... However, all is not what it seems.

... I have come to notice that you are mistaken in thinking that the two electric machines (Progressive Treatment Unit and Transidolar B225) were not as you assumed ECT machines. They are in fact muscle stimulators intended to build up strength in wasted muscles of invalid or non-ambulatory patients (not unlike the Slendertone or Ab Exercise Machines available today). Obviously much larger in size due to them not having the benefit of the miniature components and integrated circuits we take for granted these days. (http://www.simoncornwell. com/urbex/hosp/ch/e130702/96.htm)

(In)Security and Danger

What to look for once inside is but a part of the exploration process. Our research materials revealed a third important trope concerned with getting in and keeping safe once within the asylum (Table 7.4). Here we see intriguing contrasts between a present in which people are excluded from asylum sites that have, through dereliction, become unsafe places, and a past where excluded people were confined within asylums that were constructed as places of safety.

Table 7.4   (In)security and danger: sample research material

Theme	Site	URL
No entry signs	Cane Hill	http://tx.mb21.co.uk/gallery/cane-hill.php
Security	Various	See text
Decay, dereliction and vandalism	Graylingwell	http://www.28dayslater.co.uk/forums/ asylums-hospitals/60873-graylingwell- asylum-may-2011-a.html

The 'Do Not Enter' signs that encircle derelict asylum sites foster the challenge of accessing an embargoed place, to which entry is restricted due to health and safety concerns, the risk of fire or further damage or more simply as an exercise of property rights (Figure 7.5). Entry can be seen as act of subversion, impelled by a lack of human presence and ambiguities of ownership and authority as well as curiosity and a desire to record dereliction. At the same time, it is also a statement about alternative rights and freedoms. In this way there are echoes of the distinctions that Pinder (2005) draws between 'writing the city' and 'rights to the city'. The risk of entry is embraced by explorers for whom, we speculate, the intrigue and danger are something of a palliative to the relentless sameness of city spaces. In this sense, asylums are 'authentic' places (Relph, 1976), sites with a past as well as a differently dangerous present. They are accessible (given their location frequently close to or in cities) yet paradoxically distant and rendered excitingly Other through their abandonment and stigmatised reputation:

10–31–04 Arrest Log Sunday Three teens were arrested after they illegally went onto the Danvers State Hospital property, which has been closed since the early 1990s. Eric X 19, Bryan X, 17, and Andrew X 19, were all arrested and charged with trespassing by Patrolman Scott Frost. Security officers paid by the state found the teens on the hospital property around 11 p.m. Frost said the teens said they had seen the horror movie Session 9, which was filmed in part at Danvers State, and wanted to check out the property themselves. (http://www. danversstateinsaneasylum.com/2004.html)

Even at less celebrated sites these themes are clearly evident. At Graylingwell, entrance is a challenge that has to be addressed in a fashion reminiscent of classic mountaineering, requiring multiple attempts and the services of guides with prior experience of the site:

Figure 7.5   Security Fencing at Cane Hill, Croydon, UK

I'd originally attempted to get in here a few months ago with AnotherMansCause but we set off some PIRs [personnel infrared detectors; devices to identify movement in forbidden spaces] and got caught by two security guards. Cue a return trip with Olz9181 and Robbiekhan who had previously got in but had their explore cut short by security. (http://www.28dayslater.co.uk/forums/asylums-hospitals/58773-graylingwell-asylum-chichester-march-2011-a.html)

At the same time there is also a paradoxical recognition that security plays a role in preserving the challenge of entry, maintaining a blend of preservation and dereliction, and ensuring a sense of being alone in uncharted territory:

> The security guys have done a good job looking after Graylingwell as many will know, so it's not as badly chavved up [smashed, graffitied] as many other places. (http://www.28dayslater.co.uk/forums/asylums-hospitals/58773-graylingwell-asylum-chichester-march-2011-a.html)

> Graylingwell has always been that place that has been 'there' but not really seen, only a few very dedicated bastards have made it in over the last few years. Climbing was the only way. But now, even with conversion slowly moving through the site, it still stumps people. (http://www.oblivionstate.com/forum/showthread.php/4710-Graylingwell-(Visited-2011-)-2013)

> At no point in the past four or five years had we, or anybody to my knowledge, been able to extensively explore Graylingwell's interiors. There had been a couple of occurrences of folks being 'toured' around certain areas, explorers had accessed isolated rooms and outlying buildings, but security on site was and still is such that it defeats you in both the physical and mental realm. (http://www.derelictplaces.co.uk/main/showthread.php?t=3294)

> Graylingwell had been a nemesis for years, nailed windows and nervy security guards kept even the best explorers at bay and only a few made it through. Even then, they found the interior to be locked down and their movements limited. It sat quietly on the edge of the collective conscience, taunting us from behind a simple chain link fence. (http://thetimechamber.co.uk/beta/blog/asylum-hindsight)

Once within the asylum, people with mental health problems might have expected to find themselves in a safe but secure therapeutic setting conducive to recovery, though the reality was often very different in the latter years of the asylum era. Today, decay, dereliction and vandalism mean that explorers encounter a very different setting: one where the environment is frequently far from safe (Figure 7.6). This state reflects, coincidentally, a further factor related to the closure of asylums – the high costs of maintaining asylum buildings (see Chapter 1). Dangers within the derelict asylum can, on occasion, verge on the sublime and photogenic: the encroachment of vegetation, the entanglements of invasive plants and the ravages of arson and weather. More often there is simply danger that can harm the explorer, both overt in the form of rotting floors and less obvious as in the case of guano deposits and asbestos:

> After a quick look around the mortuary, trying the doors and windows, we went back down to the boiler house to find a way in. Ignoring the asbestos signs, we found that some of the boards over the windows were loose. After weighing up

the pros and cons, I went in, using a fleece to cover my mouth as a poor way of stopping asbestos. (http://www.simoncornwell.com/urbex/projects/ch/websites/urbex/old/9.htm)

Mostly just a bunch of empty rooms and a fuckload of pigeons (both dead and alive) everywhere. By the time I got up to the third floor they really were a little too close for comfort so I didn't hang around for very long. (http://www.uer.ca/locations/viewgal.asp?locid=24935&galid=19991)

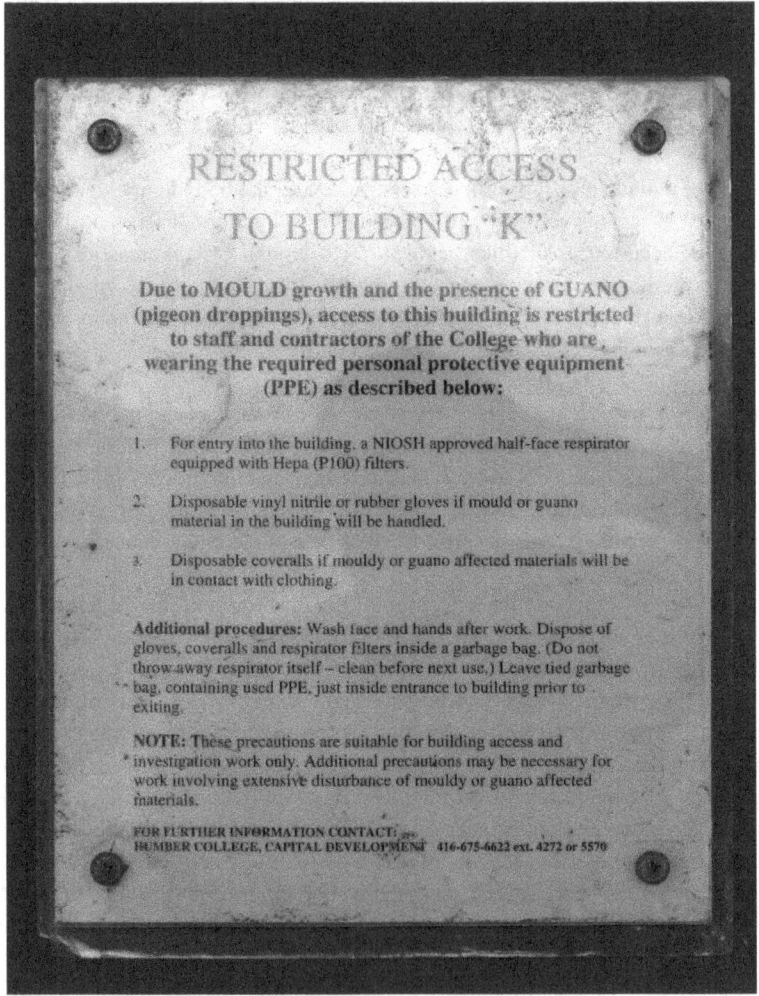

Figure 7.6   Risk and Danger at the Former Lakeshore Asylum, Toronto, Canada

Whereas past residents may have stayed in the asylum for months or even years, the explorer's stay is thus typically short. They are discoverers who move through the asylum, seeking destinations, noting details, attaining goals and moving on. Though some may stay overnight, most are day visitors, albeit sometimes returnees. Their visits are brief encounters, fraught not only with possibilities of discovery but also the prospect of harm, and accompanied by olfactory and other sensory reminders of decay that create characteristic smellscapes (Porteous, 1985) and a heightened sensual response (Rodaway, 1994) to the forbidden:

> ... barriers made movement around the complex difficult, ... blocks also stultified the movement of fresh air, creating a wet, humid and stilted atmosphere, exacerbated by overflowing blocked gutters and airbricks choked with Buddleia growth. It was a perfect culture for wood eating fungi, and Cane Hill began to slowly cook and rot. (http://www.simoncornwell.com/urbex/projects/ch/ue/cam1.htm)

In seeking out such derelict asylum spaces, in a state of alertness to danger and eviction, explorers are immersing themselves in highly sensory environments. It is to a further aspect of this sensory experience that we now turn.

Spectral moments

In his insightful work on W.G. Sebald, John Wylie (2007, p. 5) writes of worlds that are "peopled by ghosts and by the haunted: the displaced, traumatised and exiled". He goes on to point out how Sebald's geographies are not only about "places and people haunted by the insistent ghosts of the past" but also about places that are simultaneously unsettling and unsettled (Wylie, 2007). The fourth theme that we identify from the recorded encounters of urban explorers with former psychiatric asylums engages with what has been termed the spectral turn (Holloway and Kneale, 2008; Luckhurst, 2002). We use the word 'spectral' here to signify a running theme within our research material that references the shadowy presences of past inhabitation. These presences take two forms (Table 7.5). First, there are places within the derelict asylum that might be termed spectral destinations. In contrast to destinations that are sought for their architectural or psychiatric significance, these are places that evoke the past collective presence of former asylum staff and residents. Second, there are spectral manifestations that deal with the more explicit and popular understanding of the word spectral, engaging with the stereotypical imaginary of the haunted asylum and the ghostly presences reputed to drift through the corridors of some such sites.

Table 7.5   Spectral moments: sample research material

Theme	Site	URL
Destinations		
Shops	Graylingwell	http://wevsky.blogspot.co.nz/2011/05/graylingwell-asylum-may-2011.html
Theatres, Chapels, Swimming Pools	Graylingwell	http://www.uer.ca/forum_showthread_archive.asp?fid=90&threadid=51770
Storerooms, Kitchens, Wards	Cane Hill	http://www.urbexforums.com/showthread.php/787-Cane-Hill-Asylum-(July-08) http://www.forlornbritain.co.uk/canehill.php
Manifestations		
'Ghosts'	Danvers State Lakeshore	http://www.opacity.us/image3354_backside_panorama.htm http://www.opacity.us/article9_all_souls_
	Kingseat	that_haunt_this_site_can_expect_arrest.htm)
	Carrington	http://www.torontoghosts.org/index.php?/20080815194/The-Former-City-of-Etobicoke/The-Former-Lakeshore-Psychiatric-Hospital.html http://www.asylumbythelake.com/haunted-or-not/ http://torontoist.com/2012/10/torontos-haunted-hot-spots-or-should-that-be-cold-spots/ http://www.nzherald.co.nz/lifestyle/news/article.cfm?c_id=6&objectid=10806805 http://nzghosts.freeforums.org/kingseat-psychiatric-hospital-t624.html). http://www.usu.co.nz/node/2881

Spectral destinations come alive through imagining a past collective presence. They are former places of congregation, once peopled but now deserted. Some are mundane, like shops and post offices, former places of transaction and congregation within the asylum where daily lives were lived but which were also portholes to normality and the outside world. Others like theatres, assembly halls, swimming pools (Figure7.7) and chapels are more clearly event-based rather than quotidian and symbolise communal spectacle, recreation, and ceremony, hinting at the moments that fractured and enlivened the routine of the asylum.

Sites of communal spectacle can also act as architectural destinations; chapels have, for example, provided some of the most celebrated instances of the redevelopment of asylum buildings and they are also a frequent subject of preservation via heritage legislation. After closure, their retention facilitates continued public engagement with the former use. Facilities that were once places where asylum residents could imagine alternative futures become places where

Figure 7.7   Abandoned Swimming Pool, Kingseat Auckland, New Zealand

people unfamiliar with asylums can imagine a lost world of troubled souls. What makes relatively mundane parts of asylum sites into spectral destinations is the absence of the peopling that was once their raison d'être:

> … the chapel is by far the best pic, proper haunting. you can imagine loads
> of nutters in there being preached salvation to. (http://forum.breakbeat.co.uk/
> printable.aspx?m=1970484703)

What we see in these spectral destinations is a haunting absence that valorises the marginality of people with mental health problems in a way akin to that explored by Cameron in her investigation of indigenous spectral presences in Canadian landscapes (Cameron, 2008). Former psychiatric asylums are significant and interesting by virtue of what-is-not-there: the vanished people, both staff and patients, that made asylums into living communities and whose absence now enhances the impact of dereliction. Theatres, cinemas and the like are somehow out-of-place and unexpected within a carceral environment. They convey the scale of the asylum yet, in their dereliction, transmit an uncanny stillness and sense of being frozen in time. This interpretation resonates on the one hand with Conradson's ideas about organisational space, stillness and therapeutic environments (Conradson, 2011). On the other hand, there are powerful suggestions of even more toxic carceral environments. These are at their starkest in mundane ward, storeroom and kitchen settings where abandoned records, clothing, foodstuffs, beds and mobility equipment recall similar displays of past presences at Nazi death camps (Charlesworth, 1994; Stenning et al., 2008)

In turning to the encounter between the urban explorer and our notion of spectral manifestations, we engage more directly with the stigmatised stereotype of the psychiatric asylum. For some explorers there is a clear sense in which the asylum evokes ghostly presences that are, in some sense, 'real':

> I was in the Asylum at Lakeshore (Etobicoke), it gave me some very bad vibes, you kind of absorb a little of the fear and hate and anger and confusion that the old patients left behind like a residue. Anyone even slightly sensitive will tell you that, and add that you frequently feel like you're watched, or that any second you will "see" the place as it was. That is one creeeepy place. (http://www. torontoghosts.org/index.php?/20080815194/The-Former-City-of-Etobicoke/ The-Former-Lakeshore-Psychiatric-Hospital.html)

The significant word in this narrative is 'residue' – the sense that a presence has been left behind by the former asylum residents. It is of course such feelings that lie behind the aforementioned reframing of Kingseat as Spookers. They also provide the underpinning plot device for the (re)presentation of the asylum in horror movies. Some explorers explicitly recognise the stigmatising inevitability present in these discourses:

> Tunnels, abandoned old buildings, notices forbidding entrance, Gothic Revival architecture, so-called incurable cases, and even the name itself, "psychiatric hospital/asylum", have been enough to fuel ridiculous rumours and outlandish urban legends. (http://www.asylumbythelake.com/haunted-or-not/).

While reputed hauntings can invoke fear, as the foregoing comment suggests, scepticism is another response. In this sense, spectral manifestations are a means of making sense of, challenging and problematising the unusual (Maddern, 2008). They are a sensing of the past and an affective manifestation of emotion. Holloway further suggests that infrastructural remains such as abandoned buildings specifically engender supernatural possibilities (Holloway, 2010; Holloway and Kneale, 2008). At the same time there is also, for us, a sense in which spectral manifestations enable explorers to empathise with the discredited past of the asylum. With these perspectives in mind we can move from the imagining of spectral manifestation to tales of actual ghostly encounters.

Some ghostly encounters within our research material were 'classic' in their appropriation of filmic imagery of Otherness. Thus we have a ghostly woman in the basement at Lakeshore and spinning waste bins, while Cane Hill in England was reputedly haunted by the ghost of a former superintendent and bedevilled by ghostly mists and orbs. Others are constructed by popular media. Kingseat is reputedly New Zealand's number one haunted location (Stuff, 2010) but is not alone. The student newspaper at Unitec, on the former Carrington Hospital site (see Chapter 5), indicates that:

> Few students who study at Unitec would be unaware of the rumours of ghost sightings in Building One, the main building of the original Asylum. Some ghost-busters attribute the prevalence of sightings at these institutes to the high number of deaths in one location; others fault the fear and stigma surrounding mental health itself (http://www.usu.co.nz/node/2881)

While the latter statement veers towards seeking an explanation for spectral presences and takes us back to the notion of stigma, it was accompanied by a lengthy listing of allegedly haunted locales on the Carrington site.

Urban explorers and ghost hunters approach spectral manifestations in former asylums with a mix of credulity and rationality. Some are undoubtedly eager to witness something strange, reinforcing prior stereotypes. Many however develop more rational accounts. For security personnel at Carrington, hauntings are urban legends that circulate within a credulous student community and are bolstered by " ... eerie noises [that] are the result of air entering joints of old windows, and creaking and banging ... caused by the old iron heating systems" (http://www. usu.co.nz/node/2881). Others go further and offer outright denials of spectral presences: "I've read up on the history of the buildings and "sightings" that others have reported. I am fortunate enough to have had access to the school's tunnels for 3 years now. I can quite honestly say that I haven't experienced a thing down there!" (http://www.torontoghosts.org/index.php?/20080815194/The-Former-City-of-Etobicoke/The-Former-Lakeshore-Psychiatric-Hospital/All-Pages.html).

Our research material suggests that time plays a key part in the genesis and prevalence of these spectral manifestations. In a rejoinder to a popular notion linking haunting to the unquiet spirits of the dead, a former Lakeshore employee noted that the asylum had not been known as a haunted location while it was in operation though death was commonplace. Moreover: "It is obvious that a number of people died there ... but a lot of people die in general hospitals as well, and these don't have the reputation for being haunted". For this employee, dereliction and abandonment were temporally and spatially associated with haunting: "'ghosts' started to appear after the closure, when the buildings stood empty and, being easily susceptible to vandalism and homeless encampment, they began to deteriorate". (http://www.asylumbythelake.com/). The implication is that ghosts are more likely to be the inevitable noises associated with empty buildings rather than the spectral presences so eagerly sought by some.

We end this section with one further observation on the importance of both time and space in interpreting the spectral geography of the asylum. Some members of the ghost hunting community pursue their interest using technological paraphernalia. They claim to detect ghostly presences via fluctuations in magnetic fields, air quality, radiation levels or temperature, and by using infra-red movement sensors and various enhanced forms of photography. Suspending (dis)belief in this pseudoscience, we note the report on one ghost hunting forum of the impact on its host institution of Spookers, the horror experience at the former Kingseat asylum that prompted this part of our investigation. It appears that, during the period of dereliction prior to the opening of Spookers, electromagnetic field readings at Kingseat " … were anywhere from 0.4 to 18 miligauss and now it's a steady 1.5 everywhere. Spookers spooked the spooks! … Spookers brought an end to quite a few things at Kingseat. I think you'll find that anything that may have been there will now have gone. The disturbances to the electro-magnetic field will literally have destroyed their surroundings. Nobody likes noisey [sic] neighbours" (http://nzghosts.freeforums.org/kingseat-psychiatric-hospital-t624.html). While derelict asylums may be amenable to the development of spectral narratives, it thus appears that redevelopment appropriating asylum and spectral stereotypes can damage established spectral manifestations. The commodification of horror subdues some of the '(super)natural' mystique of place.

Conclusion

As noted earlier in this chapter, there is an emerging and fascinating geographical literature on urban exploration. Here our interest has not been with urban exploration *per se*. Rather, we have used the narratives of urban explorers and allied individuals with an interest in derelict asylums to identify the underlying tropes that characterise their narratives about the derelict asylum. An extended metaphor about exploration has been used to link these tropes, moving from considering physical destinations towards an examination of destinations that concern 'landscapes of the mind' (Porteous, 1990). Throughout, it has been evident that the exploration of former psychiatric asylums is influenced heavily by prevailing societal stereotypes about mental health and especially the asylum treatment modality. We see this in the physical locations that are explored, in the psychiatric signifiers that are sought out, in encounters with security and danger and in the spectral re-peopling of the abandoned sites.

Given the pervasive demonisation of the asylum it is perhaps inevitable that these tropes emerge from the narratives of explorers. What is distinctive however is that, alongside the stigma and demonisation of the asylum past, urban explorers also evidence curiosity, fascination and even, on occasion, sensitivity. We contend that the accounts that we have presented offer a particular form of nostalgia, one in which an unpleasant and uncomfortable past is sought as a yardstick by which to appreciate the present. To this extent we can link our conclusion to the thinking of Bonnett and Alexander (2013) for whom urban nostalgia can be 'productive' and related to a sense of loss. Seeing the asylum and being in the asylum allows the present-day explorer-visitor to appreciate it as a space of abjection, where past traumas were enacted. Equally, the accounts of explorers also recast this abjection, particularly through photography, into an aesthetic experience of the discredited asylum past that hints at a sense of loss, not only for those detained but also for the buildings themselves.

We argue that this performance of exploration, and indeed this envisioning of the asylum, while original in the minds of each explorer displays a high level of mimicry. The research material we examined reveals that for each asylum there are key destinations. These sought-after features amount to a virtual travel guide. The internet records of exploration constantly reinforce this guide and give it validity, framing and constraining the journey of subsequent explorers, establishing canonical accounts of the dereliction of each asylum. For explorers, the primary gaze is on the richness of the spaces of the derelict asylum framed by stereotypes of a hazily understood past. The haziness is perhaps significant: many explorers will have little direct knowledge of asylum care given the lengthening time since the asylums closed. Imagination is central to interpretation, yet even here there are common threads drawn from popular literature and media accounts.

While there are broad connections between the foregoing observations and the ideas of strategic forgetting and selective remembrance introduced in Chapter 2 and explored in depth in Chapters 5 and 6, we note important differences. In the case of repurposing for tertiary education (Chapter 5) and housing (Chapter 6), we outlined the systematic role of institutional action in shaping strategic forgetting and selected remembrance. In essence, they were bundled within marketing and rebranding strategies that sought to distance the new use from the stigmatised past while valorising the historic legacy represented in buildings and landcapes. In contrast to this corporate view, which tacitly renders previous practices and users invisible, urban explorers value the connection between buildings and their past uses and occupants. The sense of intrigue which drives the activities of explorers and the dark romantic vision they seek to recover is sufficiently strong to lead them into sites of both risk and reward. In doing so, they see beyond the architectural merit of buildings and the beauty of landscapes to recover remnant evidence of the stigmatised past – albeit one shaped by popular culture and the legacy of previous practitioners of this clandestine activity.

While we have focused on selected case study asylums, we have no reason to doubt that similar tropes would not emerge had we chosen other sites. Using web-based material has however posed challenges. The material can be ephemeral. In some instances, images and accounts we had drawn on disappeared from the internet between the research and writing of this chapter. We will return to this important methodological issue in Chapter 8. For now, we recognise that some of the tropes that we identify might apply equally to other derelict buildings. Indeed explorers do not specialise exclusively in former asylums so our thematic analysis may well have wider applicability. Above all though we are conscious that we have focused on material produced by a group of people who, we contend, can offer particular insights. Specifically, their incursions into asylum space are generally uncoloured by belonging or past memories other than those held collectively by society at large. To this end their 'remembrances' stand in contrast to those of former asylum residents or staff, the other main constituencies writing on closed and sometimes derelict asylums. Urban explorers are seldom interested in asylums *per se*; for them the challenge is getting in, staying safe and undisturbed, and recording their visit rather than recovering a personal experiential understanding of the asylum. Following Pinder (2005), we see explorers as people (mainly but by no means exclusively young and male (Garrett and Hawkins, 2013; Mott and Roberts, 2014)) whose guerrilla achievements are undertaken initially out of a sense of entitlement but who also 'give back' to society through exhibiting their exploits in the public domain of the web. Their accounts are often distinguished by a brutal lack of sentimentality. Staff and patients are interested in people and the past; urban explorers provide us with accounts that focus on buildings, sites, the present and dereliction. To this extent, in contrast to staff or patient accounts, explorers reflect what Bonnett (2009) has seen as 'revolutionary psychogeography' in which a traditional nostalgia is eschewed.

We have noted elsewhere that asylum closure seldom evoked the protest or active resistance seen when general hospitals or schools have been threatened with closure (Kearns et al., 2009; Moon and Brown, 2001). We see asylum sites left derelict in the wake of policy change as spaces of cultural and social loss, generative of both disdain and particular expressions of nostalgia. While any abandoned building potentially raises curiosity about past use and occupancy, the particular mystique of madness and the long shadow of carceral care mean the abandoned asylum is an especially unsettling space. The speed of the removal of former occupants and the banishing of now-vilified practices has left a legacy comprising a potent mix of fear, disgust and intrigue, suggestive of a site-specific emotional geography. We have argued here that these remnant spaces of the asylum have spectral qualities, in the sense that there are material presences of the past ranging from architectural forms to hospital furniture and medical paraphernalia. The performance of exploration, and the records its practitioners leave in the public domain, challenge our sense of normality, hence it is unsurprising that the interest that these material prompts generate often leads to rumours of paranormal presences in these spaces of radical absence.

Michel de Certeau suggests, that "there is no place that is not haunted by many different spirits hidden there in silence" (de Certeau, 1984, p. 108). We see the derelict psychiatric asylum in this frame. The kaleidoscopic assemblage of material remnants evident in the accounts of urban explorers visiting former asylums serve as an *aide-mémoire*. They keep alive the past and remind us of the asylum. While derelict, asylum sites are places in waiting, simultaneously sites of play, danger and discovery, they are also liminal sites positioned on the edge of current experience where the asylum past is receding into the distance but where, for the moment it can be encountered in the present.

Chapter 8

The Future Place of Past Asylums

In this chapter we reflect on our research programme on the persistence of the idea of asylum and the re-use of former asylum sites and buildings. We organise our observations in two major sections. The first of these deals with our research journey. We begin with a reflection on our overall approach and the constituent methods we adopted. We then offer suggestions for future directions in which the research could be taken. In a second major section we re-visit the pivotal questions posed at the beginning of the book, asking what has happened to the idea of asylum and what has happened to the sites and buildings of former asylums? We reconsider the transcendence of policy in light of our case studies. We then offer summative insights on the dynamics of memorialisation and conclude with a reconsideration of the power of stigma.

The Research Journey

Reflecting on our Approach

The previous chapters have been built around strategically-chosen case studies. A case-focussed approach allowed us to anchor and illustrate our key concepts (Yin, 1989) and to consider processes such as stigma, remembrance and memorialisation without separating them from the contexts of their unfolding. As Flyvbjerg (2006, p. 219) argues, "social science may be strengthened by the execution of a greater number of good case studies". Indeed, while the few prior studies of asylum repurposing have largely relied on survey-based stocktakes (see, for example, Chaplin and Peters, 2003), we see our approach as having the considerable advantage of iteratively developing a standard observational approach and allowing, in most cases, revisits over time. A survey approach would invariably rely on second-hand observations and would raise questions of reliability as respondents would vary in terms of their positionality and knowledge of the site. Indeed, in our experience it would be challenging to establish who exactly the appropriate and available respondents would be; in some cases it was evident that no one was taking responsibility for the site of a closed asylum, or alternatively the imperative of corporate confidentiality meant that there was a filtering of information lest marketing possibilities be compromised. Despite the inevitable selectivity of our case study approach, in one country (New Zealand) we were able to visit all sites and in some cases multiple times. This opportunity permitted us to assemble a complete data set that served as a window into the changing fates of former asylum sites and buildings (Kearns et al., 2012). Within our suite of methods, we adopted a 'forensic approach', through which we searched for hidden aspects of memorialisation on site and in text. We initially took this detective work for granted but then began to see it as a distinctive and valued part of the research. We were able to triangulate information focussed on single sites. This process involved field visits with opportunistic conversations with local residents and other informants, some of whom had a close relationship to the former asylum through employment or having a family member as a patient. A further dimension of triangulation was the subsequent mining of the media record, thus creating a temporal window into past processes leading to the current state of the site.

We see our experience of field observation as allowing the building of a four-dimensional understanding of each site. Clearly any visit is coloured by first impressions of the layout of the grounds and the character of the buildings on the site gained through walking through the grounds. This process amounts to encountering the expected three dimensions of any built environment. A fourth dimension was accessed through sequential visits which allowed the addition of the critical dimension of time; the witnessing of change in the appearance and function of former asylums. We organised most visits such that at least two team members were present. This arrangement allowed us to maximise critical observation at the sites and develop a deep interpretive reflection in the days following. These visits were informed by cultural-geographic perspectives in which the landscape is 'read' for its absences as well as presences. We sought to collate and organise what Anderson (2009) calls 'traces' of past use and human occupation.

To extend the reach of our research, we consulted media reports and websites. The former allowed us to assemble sequences of events and influences that shaped the fate of particular asylums. The latter, websites, offered two further vantage points: first, a window into official attempts to brand and market private asylums and repurposed asylum sites; and second, a record of the impressions and achievements of self-styled 'explorers' who knowingly venture into abandoned asylum buildings. We expected the *ad hoc* character of these sites to result in a certain level of frustration with data disappearing from sight/site. What was less anticipated was the rapid change in the way successor-users changed the information they offered visitors. By way of example, on the Unitec website there was initially a basic history of the grounds and details of its former existence as Carrington Hospital. Within three years, this material was removed from the Unitec website such that the entire site had no official reference to a past use which had persisted until just over two decades ago. We can conclude that, in the quest to rebrand new uses, the half-life of information can sometimes be very short. In contrast, others sites may continue to host information that endures long after it is outdated. By way of illustration, the web records of explorations of derelict sites cover several former asylums where demolition or redevelopment has now taken place. The type-site of Cane Hill is perhaps the prime example.

Some comment on our experience of field observation is warranted. Others have pointed out the obvious: former asylum sites are locations that are frequently imbued with an almost palpable atmosphere arising as much from knowledge of past practices and lives as from the persistence of evidence of those practices (Payne and Sacks, 2009). While asylum practices have been vilified and many lives ended in sadness and despair, for many residents these places were warmly regarded and generated a sense of belonging, consistent with the original concept of asylum. Hence, the asylum can be generalised as a site replete with the legacy of extreme valences of emotion (see Davidson et al., 2005). For us as researchers, this situation translated to encounters with sites in a more-than-material manner. By this we mean that our fieldwork was affected by past presences and practices. By way of example, at Tokanui, (New Zealand), discussion of a particular part of the site cued a memory shared earlier with one of the authors by an academic colleague who grew up in the area. She recounted the impact on her of being taken on a school trip to the hospital in its heyday and 'viewing' a number of 'mentally handicapped' babies and toddlers. The profound effect on her was transferred to us as observers of the abandoned site decades later. As a second example, on a return visit to Kingseat south of Auckland, we made the assumption that the site was still abandoned apart from the horror-themed business, Spookers, located in a far corner of the estate. It was a considerable shock when suddenly we sighted lines of washing strung out to dry between decrepit wards. Subsequent investigation of the site revealed a large number of unofficial paying tenants who were occupying unmodified decaying buildings. This was an unsettling experience; we were wandering in what felt like and looked like an abandoned site, but it was actually a *de facto* neighbourhood. The asylum was being covertly resettled, a process otherwise invisible other than by direct observation. We could only wonder how these tenants themselves felt about their ephemeral living quarters in the once-asylum. This 'overflow' accommodation was subsequently observed at two other New Zealand sites, which speaks to the role of closed asylums in offering marginal accommodation close to cities for those otherwise excluded in overheated rental markets.

Future Directions

In terms of possible future research directions, we note possibilities for conducting case studies in countries other than the three on which we chose to focus. Such an expansion of scope would almost certainly provide interesting counterpoints to our results and provide additional feedback on the approach and the suite of methods we employed. In terms of the persistence of the idea of asylum, it would be particularly useful to see whether, and to what extent, the dependence of private provisioning of residential care on public funding has been a universal phenomenon. We have made brief reference to other potential national case studies and contend that an examination of closure narratives would be useful in countries such as France (where, as noted in Chapter 4, there remains a degree of preference for residential forms of care), the US (where there are particularly fine examples of asylum heritage, noting our coverage of Kalamazoo in Chapter 5 and

Danvers State in Chapter 7), or Germany (where there remains a long shadow of the eugenics movement widely practiced in Third Reich asylums) .

With respect to the re-use of former asylum sites and buildings, we also see merit in further visits to our case study sites, especially with an eye to the fate of derelict sites and those where re-development trajectories were unclear at the time of our case study. Our stocktake of developments in New Zealand (Kearns et al., 2012) provides a template for such an update and the brief discussion in Chapter 4 points to interesting developments in Ontario as residential mental health care has reappeared on the sites of several former asylums. It would be interesting to know how far this resurgence will go and whether these developments can be distanced from the shadow of the Victorian asylum, the remains of which co-exist on some sites. A parallel question in the UK relates to whether the appetite for the conversion of former asylum grounds and buildings into housing estates will continue.

Additionally, we see merit in investigating the fate of other custodial 19th Century institutions. Some of these have strong similarities in their origin to the psychiatric asylum. Indeed, the Victorians in particular developed institutional responses to a range of social and health problems. Sanatoria, orphanages, homes for 'mentally-handicapped' children and adults or 'colonies' for people with physical disabilities, workhouses, residential schools for aboriginal people (in Canada, the United States and Australia), Magdalene Laundries (in Ireland) and prisons come to mind as interesting candidates. Such investigations hold the potential to reveal what is general about the re-use of redundant infrastructure and what is distinctive about particular institutional networks. For instance, to what extent do stigma or histories of abuse play a role in colouring discussions of the re-use of any outmoded institutional site and buildings? Is the psychiatric asylum unique in its layered stigma, or is it the highly visible portion of a veritable iceberg of institutional sites made difficult to re-use by the stigma associated with previous uses, their institutions and the people to whom they catered?

Policy, Remembrance and Re-use

In looking back across the chapters in which we examined the idea of the four fates, we organise our concluding thoughts around three themes. In each, we reflect on the key ideas and interpretive framework introduced in Chapter 2. The first of these themes considers the transcendence of policy, both within and outside mental health care, in determining the survival of the idea of the asylum and in shaping considerations of the re-use of former sites and buildings. The second theme explores the complex dynamics of the memorialisation and remembrance of the asylum. Third, we reconsider the key construct of stigma which we have argued shaped not only the lives of individuals but also added a potent valence to the reputation of sites of former treatment. Within these discussions we focus not only on what we consider to be key insights, but also highlight those that run counter to our original expectations. These insights include the dependence of many remaining private sector psychiatric asylums on funding from the public sector, the re-emergence of residential care as a public sector treatment modality and the hybridity of the four fates of former asylum sites.

The Transcendence of Policy

The survival of the idea of asylum, whether in the form of private psychiatric care or the persistence of services on former public asylum sites, is permeated by indications of the profound influence of policy. Initially, and as noted in Chapter 2, it was deinstitutionalisation that set the agenda for change; the psychiatric asylum was deemed to be an outdated modality of care and redundant to the needs of contemporary society. In associated discourse, the limitations (and often the perceived evils) of institutional care were subject to close scrutiny, while those of the favoured option of community care were ignored or characterised as 'teething troubles' to be overcome as new services were developed. The emergence of neo-liberal imperatives for restructuring created additional incentives for closing large institutions weighed down by substantial deferred maintenance bills and the prospect of escalating operating costs provided a 'hard edge' to the process of deinstitutionalisation.

Reaction to pressure for deinstitutionalisation varied between the private and public sectors, from place to place, and over time. In the private sector, key distinctions need to be made with respect to institutional reactions prior to and after the emergence of restructuring as a force shaping the retreat from residential care. With respect to the earlier of these phases, Homewood is illustrative of the need for private asylums to find a new niche, in both a legislative and market sense, in an era of community care (see Chapter 3). The survival of Homewood was accomplished through a combination of public relations (in which the hospital traded heavily on its reputation for innovation) and the acceptance of hybridity. While retaining its status as a private-sector hospital-based institution, Homewood agreed to dedicate a significant portion of its capacity to public sector (and funded) in-patients and took the initiative to convert part of its capacity to out-patient and community care. Our exemplar institution in New Zealand, Ashburn Hall, was similarly challenged by the move toward community care as the new norm, but in this case financial survival involved, at least for a period, public ownership under the auspices of a university medical school (see Chapter 3).

The full flowering of restructuring brought with it a more favourable view of private provisioning. The growing acceptance of the contracting out of a range of mainly non-medical services in hospitals, the privatisation of many stand-alone diagnostic services and the emergence of high profile entrepreneurs in the residential care of the elderly (increasingly seen as an industry rather than a social service) made 'institutional survivors' like Homewood and Ashburn less an anomaly and more an example of new possibilities. The Priory Group (and especially its founder Dr. Chai Patel), our UK exemplar, was very much the poster child of the entrepreneurship associated with 'new' private provision. The Priory Group deployed a whole range of public relations and marketing strategies to establish the place of its residentially-based mental health services within the UK and abroad, including advertising replete with images reminiscent of therapeutic landscape (see Chapter 3). However, the reality of developments in private psychiatric care was more modest than the rhetoric through which they were often promoted. For example, in the Homewood case, significant growth occurred in non-residential services such as employee assistance programmes and in residential care for the elderly, but the residential (hospital) part of the enterprise remained virtually static in size and consistently dependent on state funding. The recent acquisition of Homewood by the Schlegel Group, a large private provider of residential care for the elderly, has guaranteed that this de-emphasising of the once-core hospital function will continue (see Chapter 3). In the UK, the financial trajectory of the Priory Group, its several recent changes of ownership and leadership, and the ongoing importance of the public sector as paymaster underscores a result that runs counter to our initial expectation. While the private sector has been a site where the building blocks of the notion of asylum have been preserved, it has generally done so as the handmaiden of the public sector. As noted in Chapter 3, the continuation of this financial support of the private residential option is made problematic by the re-emergence of public residential care in some jurisdictions, most obviously in Ontario, Canada (see Chapter 4).

The survival of the idea of asylum can also be seen in the persistence of public mental health care services on former asylum sites, often, but not always, in former buildings. These sites of continued use are not necessarily still physically obscured by walls, fencing or landscaping. What was once comfortably away from the public gaze may now be re-purposed as a community resource (see Chapter 4). Echoes of the separated asylum persist, however, with forensic units that have survived, even in situations where the greater part of sites has been re-developed for housing (e.g., Knowle; see Chapter 6) or as a tertiary education facility (e.g., Carrington; see Chapter 5). The obscuring of the survival of residential care on former asylum sites through re-naming has also been important. The case of Sunnyside in Christchurch (New Zealand) is emblematic of this strategy. As noted in Chapter 6, Sunnyside was closed at the 'stroke of a pen', with a range of services (including residential care) being 'transferred' to the successor hospital, Hillmorton, operating on the same site in the same buildings. This distancing from the asylum legacy was also evident in Ontario and the UK where mental health services continue to be offered on (parts of) several former asylum sites in 'mental health villages' or 'clinics' in new buildings that are invariably modest in size and design compared to their asylum predecessors. It is notable that the third generation facilities being built and opened on these same sites derive their naming from therapeutic euphemisms or their modest and relatively short-lived predecessors rather than the earlier, impressive and long-lived, asylums (see Chapter 4). This situation brings to mind the notion of 'strategic forgetting', which we will highlight in the following section.

In those many instances in which asylums were closed, trajectories toward decommissioning were influenced by various policies and trends at the national level and mediated by local conditions and actors, each of which has a distinct and significant time dimension. In Chapter 5 we noted the complementarity of the timelines for the closure of psychiatric asylums and the expansionary trend in post-secondary education common to our three case study countries. Carrington in Auckland, where patients awaiting transfer in the early 1990s almost literally rubbed shoulders with incoming students, serves as an excellent example of such serendipity. However, such a conversion from hospital to college campus was not inevitable, but rather was made likely by the favourable location of Carrington and the willingness of key actors at what is now Unitec and elsewhere to make it happen. In our other case study in Chapter 5, Lakeshore in Toronto, the locational advantage of the site was similar to the Carrington case, but here the journey to re-use was considerably more protracted and contentious and even more demanding of strong leadership from senior administrators at Humber College.

Time was an important dimension of each of the journeys to re-use mentioned above and, as in any journey, there was also a spatial dimension. In Chapter 7, in our discussion of the dereliction fate we noted a particular interaction between the temporal and spatial dimensions. Isolated asylums in areas of low growth attracted little or no interest from developers and their buildings fell quickly into decay, thereby making re-development even less likely. However, the passage of time can be positive. For example economic conditions may become more favourable and brownfield sites, even those encumbered with decrepit buildings, may be seen as more valuable. This is especially true of the re-development of sites for housing, where urban expansion over time may have the effect of making the 'once too distant' seem 'convenient and accessible', especially in locations of escalating housing demand (e.g., South-East England). Such reassessment of locational (dis)advantages may be associated with or even driven by policy shifts. By way of example, at the former Carrington Hospital demand for housing intensification in Auckland at large has encouraged its owners (Unitec) to propose a redevelopment of a large portion of the green space on the site.

In Chapter 6 we discussed the profound impact of heritage conservation policy and constituent national, regional and local initiatives on the re-use of former asylum sites (and the re-purposing of many of their architecturally-significant buildings) for housing. We noted that developers in the UK are attracted to these sites because of their location, size and landscape. Moreover, they have been able to afford the cost of re-purposing buildings with heritage designation and able, through strategic forgetting and selective remembrance embedded in marketing plans, to distance new developments from their stigmatised past. While this pathway might, in retrospect, be seen as almost predetermined it clearly is not. In Canada and New Zealand, heritage conservation has not been such an important force, even in situations where sites have been re-developed for housing or such re-development seriously considered (see Chapter 6). Indeed, in New Zealand there seems to be an antipathy toward the preservation of former asylum buildings, whatever their architectural significance. In this respect, we remind readers of the case of Sunnyside and the 'accidental de-listing' of its iconic administration building at the very time that its fate was being considered by the Christchurch City Council. In Ontario, while heritage conservation has not been ignored it has nonetheless been constrained by the fact that many asylum sites have been retained for the provision of mental health care services and subject to waves of re-development for that purpose. Most recently, and as described in Chapter 4, this has involved the building of new psychiatric facilities that reflect new ideas about residential care and a concurrent disposal of outdated (but heritage-listed) asylum buildings. We see the differential impact of heritage policy and policy regarding the provision of residential care in the public sector as the source of significant differences in the re-use of former asylum sites in the three case study countries. These differences carry with them implications for the memorialisation and remembrance of the past use, and it is to this that we now turn.

The dynamics of memorialisation and remembrance

The case studies presented in the preceding chapters reveal the dynamics of memorialisation and remembrance to be complex and evolutionary. Yet generalisation is possible along several lines associated with time and positionality. We regard time as a major influence upon remembrance. When newly closed, the sites of psychiatric asylums would have resonated with detailed and vivid memories for former patients, workers and those living in adjacent communities. Such intimate memories of everyday life in the asylum and its role in the community inevitably fade over time. As noted repeatedly in our case studies, as time passes and those who had direct experience of the asylum move away or pass on, remembrance would depend more and more on memorialisation. For some, a thread of memory could be reinforced through the continued opportunities for access to the park-like grounds afforded by the former asylum through specific types of reuse (e.g. tertiary education). In these instances, elements of *de facto* public ownership continue through customary rights of access for surrounding communities. Evidence of how such access is valued was presented in Chapters 5 and 6.

In earlier chapters we provided examples of memorialisation – the implanting of triggers for memory - at former asylum sites that are now used for educational and housing purposes. We noted that while explicit memorialisation can be important, in many instances it is the presence of former (now re-purposed) buildings and the persistence of former landscapes that has the more telling presence. However, while anyone can be informed by a well-designed plaque or museum display, the 'reading' of buildings and landscapes is difficult for the lay observer, especially those who did not live through the asylum era. This difficulty is exacerbated when there are conscious decisions to downplay prior use. In this respect we can learn from comparative examples of re-purposing. As we noted, in southern England, the same developer was responsible for the repurposing of a former naval establishment as the Gunwharf Quay Shopping Centre in Portsmouth and the conversion of the former psychiatric asylum at Knowle (see Chapter 6) into a residential development. In the first instance, the developer sought to capitalise on the former use by placing torpedoes and other military paraphernalia on the site as street furniture. These become, to adapt terminology of Hopkins (1990), 'temporal icons' that speak to a past set of activities. In contrast, at Knowle there is no parallel evocation of the past use. Indeed it is left strategically forgotten. Similarly in Auckland, the recent conversion of the Hobsonville air force base has seen a profusion of reminders of the military past, in the form of storyboards and even the naming of the popular eating establishment (Catalina cafe), all of which contribute to the branding of this innovative housing development (Opit and Kearns, 2014). Elsewhere in Auckland, and as noted earlier (Chapter 5), there is limited memorialisation of the former Carrington Hospital on the successor Unitec campus apart from the renovated asylum buildings themselves. In situations such as this, when there is little or no memorialisation, supplemented perhaps by strategic forgetting and selective remembrance by new owners, remembrance of the asylum will become increasingly dominated by portrayals in popular culture.

Not surprisingly, the internet is increasingly the dominant medium for remembrance of the psychiatric asylum. Some websites are supported by former residents or workers, while others are maintained by successor mental health institutions or by historical societies. However, there are strong voices, such as those interested in hauntings and in the exploration of derelict sites and buildings, which deploy an interpretive lens that is influenced more by filmic depictions (such as "Shutter Island") than by 'insider accounts' of the asylum. More often than not, these interpretations are a caricature of the former asylum, given the appearance of reality through guerrilla photography of derelict sites (see Chapter 7). For 'explorers', abandoned sites are treated as de facto museums and, in the absence of few actual museums dedicated to psychiatric history, the explorers' online visual and written records contribute a potent form of virtual remembrance. Nonetheless, the traces they leave in blog and photograph tend to represent a very particular construction of the asylum, and one which accentuates the gothic, macabre and carceral aspects of asylum life. It is the relative youth of the explorers that distances their accounts from direct memories of the asylum era and thus leads to the embracing of this contrived form of remembrance.

Such a re-grounding of popular culture imagery is especially powerful at sites such as the Auckland's Spookers horror attraction and in UK urban exploration websites. Beyond popular culture there is potential for people to be literally 'spooked' by their everyday occupation of places in which traumatic activity was undertaken. In the case of Unitec in Auckland, we find it interesting that although the institutional website no longer features any reference to the past uses as an mental hospital, nonetheless employees are annually offered the option of having their offices ritually 'cleansed' by a kaumātua (Māori elder).

With time, as derelict sites become fewer and their remaining buildings fall into terminal decay, 'controlled' forms of memorialisation may have a growing impact, whether they be sponsored by successor mental health care institutions on former asylum sites or by those who control successor uses in other sectors. This said, we can only speculate that there will be considerable commonalities in the way that strategic forgetting and selective remembrance are deployed and former asylum sites re-branded. What is more certain is that the psychiatric asylum will be remembered. In comparison with the largely invisible sites and nondescript buildings of the community care era, there will remain more traces of the former asylum in the cultural landscape and these will continue to influence collective memory, at least locally. The preservation of architectural heritage and renewed academic interest may provide some of the building blocks for a more balanced remembering of the psychiatric asylum. Another building block could be the expansion of museums, often on former hospital sites. Such museums can serve as a vehicle for controlling and correcting memory. However they invariably walk a fine line between educating and entertaining a curious public.

Conclusion

In closing, we return to one set of the foundations for our book. In Chapter 2 we identified a series of manifestations of stigma that amplify Goffman's original formulation. These were the stigma of spoiled identity (person-based stigma), the tainted and outdated practices related to care (modality-based stigma) and, thirdly, facilities that become devalued and regarded as insufficiently important to constitute heritage (facility-based stigma). Now, having completed our assessment, we can conclude that buildings and grounds do indeed accrue facility-based stigma. This stigma develops through past occupancy by stigmatised patients as well as memories of outmoded treatment practices. In this book we have noted examples of successful conversion of sites and buildings to new uses with limited remembrance of past people and practices. These conversions have often involved demolition and new building programmes. We can conclude that asylum buildings and former sites, while accruing stigma, can be at least partially 'cleansed' of this long shadow of stigma through careful redevelopment and strategic forgetting. As Viejo-Rose (2011, p. 477) suggests, "unresolved pasts are not easily put to rest". We therefore close by posing a paradox: can the asylum be adequately remembered when sites are effectively cleansed of stigma?

References

Abbott, M. and Kemp, D. (1994). New Zealand, in *Handbook on International Mental Health Policy*, edited by D. Kemp. New York: Praeger, pp. 217–52.

Anderson, J. (2009). *Understanding Cultural Geography: Places and Traces.* Oxford: Taylor & Francis.

Andrews, G. and Phillips, D. (2005). Petit bourgeois healthcare? The big small-business of private complementary medical practice. *Complementary Therapies in Clinical Practice*, 11(2), 87–104.

Anon. (1992). The Knowle Experience. Available at: http://knowle-village.org/The%20Knowle%20Experience.pdf (accessed 2 December 2014).

Anon. (2008). Graylingwell Community Planning Weekend. Available at: http://www.graylingwellchichester.com/downloads/newsletter.pdf (accessed 2 December 2014).

Anon. (N.D.a). Humber College is expanding. Available at: www.superbuild.humberc.on.ca (accessed 1 September 2009).

Anon. (N.D.b). Graylingwell Website. Available at: http://www.graylingwellchichester.com/index.htm (accessed 2 December 2014).

Arnold, K. (1996). Time heals: Making history in medical museums, in *Making Histories in Museums*, edited by G. Kavanaagh. London: Bloomsbury, pp. 15–29.

Ashburn Clinic. (2003). The Ashburn Clinic: Private Psychiatric Centre. Available at: http://www.ashburn.co.nz (accessed 2 December 2014).

Ashburn Clinic. (2005). The Ashburn Clinic. Available at: http://www.asburn.co.nz (accessed 2 December 2014).

Ashburn Clinic. (N.D.a). *Welcome to the Ashburn Clinic.* Dunedin: Ashburn Clinic.

Ashburn Clinic. (N.D.b). *About the Ashburn Clinic.* Dunendin: Ashburn Clinic.

Ashworth, A. (1975). *Stanley Royd Hospital Wakefield: 150 Years: A History.* Wakefield: Stanley Royd Hosptial.

Atkinson, D., Cooke, S. and Spooner, D. (2002). Tales from the Riverbank: Place-marketing and maritime heritages. *International Journal of Heritage Studies*, 8(1), 25–40.

Atkinson, R. and Flint, J. (2004). Fortress UK? Gated communities, the spatial revolt of the elites and time–space trajectories of segregation. *Housing Studies*, 19(6), 875–92.

Avondale-Waterview Historical Society Incorporated. (2002). Newsletter, November-December.

Azaryahu, M. (2003). RePlacing memory: The reorientation of Buchenwald. *Cultural Geographies*, 10(1), 1–20.

Azaryahu, M. and Foote, K. (2008). Historical space as narrative medium: On the configuration of spatial narratives of time at historical sites. *GeoJournal*, 73(3), 179–94.

Bade, D. (2010). Issues and tensions in island heritage management: A case study of Motuihe Island, New Zealand. *Island Studies Journal*, 5(1), 25–42.

Bagaeen, S. (2006). Brownfield sites as building blocks for sustainable urban environments: A view on international experience in redeveloping former military sites. *Urban Design International*, 11(2), 117–28.

Barc, A. (2009). Asylum by the Lake. Available at: http://www.asylumbythelake. com (accessed 1 July 2010).

Bartlett, H. and Phillips, D. (1996). Policy issues in the private sector: Examples from long-term care in the UK. *Social Science and Medicine*, 43(5), 731–7.

Baxter, J. (2008). *Māori Mental Health Needs Profile: A Review of the Evidence*. Wellington: Ministry of Health.

Bennett, L. (2011). "Bunkerology": A case study in the theory and practice of urban exploration. *Environment and Planning D: Society and Space*, 29(3), 421–34.

Berg, L. and Kearns, R. (1996). Naming as norming: 'Race', gender, and the identity politics of naming places in Aotearoa/New Zealand. *Environment and Planning D: Society and Space*, 14(1), 99–122.

Bigby, C. (2002). Ageing people with lifelong disability: Challenges for the aged care and disability sectors. *Journal of Intellectual and Development Disability*, 27(4), 231–41.

Blandy, S. (2006). Gated communities in England: Historical perspectives and current developments *Geojournal*, 66(1–2), 15–26.

Blank, R. and Burau, V. (2006). *Comparative Health Policy*. Basingstoke: Palgrave.

Bleakley, M., Holdwich, C. and Deane, R. (1991). Changing with the times: Reconfiguring a mental health facility in response to changing market conditions. *The Psychiatric Hospital*, 22(3), 123–6.

Bluglass, T. (1992). The special hospitals. *British Medical Journal*, 305(6849), 323.

Bonnett, A. (2009). The dilemmas of radical nostalgia in British psychogeography. *Theory, Culture and Society*, 26(1), 45–70.

Bonnett, A. and Alexander, C. (2013). Mobile nostalgias: Connecting visions of the urban past, present and future amongst ex-residents. *Transactions of the Institute of British Geographers*, 38(3), 391–402.

Bowden, K. (2012). Glimpses through the gates: Gentrification and the continuing histories of the Devon County Pauper Lunatic Asylum. *Housing, Theory and Society*, 29(1), 114–39.

Bowring, J. (2010). The search for spotless landscapes, in *Beyond the Scene: Landscape and Identity in Aotearoa, New Zealand*, edited by J. Stephenson, J. Ruru and M. Abbott. Dunedin: Otago University Press.

Bradley, A., Hall, A. and Harrison, M. (2002). Selling cities: Promoting new images for meetings tourism. *Cities*, 19(1), 61–70.

Brown, T. (1980). Architecture as therapy. *Archivaria*, 10, 99–124.

Brunton, W. (2003). The origins of deinstitutionalisation in New Zealand. *Health and History*, 5(2), 75–103.

Brunton, W. (2011). The Scottish influence on New Zealand psychiatry before WWII. *Immigrants and Minorities*, 29(3), 308–42.

Bryson, B. (1997). *Notes from a Small Island*. London: Harper Perennial.

Buntman, B. (2008). Tourism and tragedy: The memorial at Belzec, Poland. *International Journal of Heritage Studies*, 14(5), 422–48.

Burk, A. (2003). Private griefs, public places. *Political Geography*, 22(3), 317–33.

Busfield, J. (2011). *Mental Illness*. Cambridge: Polity Press.

Cameron, E. (2008). Indigenous spectrality and the politics of postcolonial ghost stories. *Cultural Geographies*, 15(3), 383–93. doi: 10.1177/1474474008091334.

Candlin, F. (2012). Independent museums, heritage, and the shape of museum studies. *Museum and Society*, 10(1), 28–41.

Canterbury District Health Board. (2007). *Healthfirst: Promoting a Healthy Canterbury*. Christchurch: Canterbury District Health Board.

Capital and Coast District Health Board. (2007). Central regional forensic mental health services: Draft 5 year development plan.

Castel, R. (1988). *The Regulation of Madness: The Origins of Incarceration in France*. Berkeley, CA: University of California Press.

Castel, R. (1991). From dangerousness to risk, in *The Foucault Effect: Studies in Governmentality*, edited by G. Burchell, C. Gordon and P. Miller. London: Harvester Wheatsheaf, pp. 281–98.

Chang, T. and Huang, S. (2005). Recreating place, replacing memory: Creative destruction at the Singapore River. *Asia Pacific Viewpoint*, 46(3), 267–80.

Chaplin, R. and Peters, S. (2003). Executives have taken over the asylum: The fate of 71 psychiatric hospitals. *Psychiatric Bulletin*, 27(6), 227.

Chaplow, D., Chaplow, R. and Maniapoto, W. (1993). Addressing cultural differences in institutions: Changing Health Practices in New Zealand. *Criminal Behaviour and Mental Health*, 3(4), 307–21.

Charlesworth, A. (1992). Towards a geography of the Shoah. *Journal of Historical Geography*, 18(4), 464–69.

Charlesworth, A. (1994). Contesting places of memory: The case of Auschwitz. *Environment and Planning D*, 12(5), 579–79.

Christchurch City Libraries. (2010). Sunnyside Hospital. Available at: http://christchurchcitylibraries.com/Heritage/Places/Public/Hospitals/Sunnyside (accessed 3 December 2012).

Clark, C. (2010). *Reuse of Historic Naval Hospitals*. Paper presented at the WIT Transactions on Ecology and the Environment.

Cloke, P. and Pawson, E. (2008). Memorial trees and treescape memories. *Environment and Planning D: Society and Space*, 26(1), 107–22.

Coldefy, M. (2012). L'évolution des dispositifs de soins psychiatriques en Allemagne, Angleterre, France et Italie: similitudes et divergences. *Questions d'économie de la santé*, 180, 1–8.

Coleborne, C. (2003). Remembering psychiatry's past: The psychiatric collection and its display at Porirua hospital museum, New Zealand. *Journal of Material Culture*, 8(1), 97–118.

Coleborne, C. and MacKinnon, D. (2003). *Madness in Australia: Histories, Heritage and the Asylum*. St. Lucia, Queensland: University of Queensland Press.

Coleborne, C. and MacKinnon, D. (2011). *Exhibiting Madness in Museums: Remembering Psychiatry through Collection and Display*. London: Routledge.

Conradson, D. (2005). Landscape, care and the relational self: Therapeutic encounters in rural England. *Health & Place*, 11(4), 337–48.

Conradson, D. (2011). The orchestration of feeling: Stillness, spirituality and places of retreat, in *Stillness in a Mobile World*, edited by D. Bissell and G. Fuller. London: Routledge, pp. 71–86.

Cook, I. (2004). Waterfront regeneration, gentrification and the entrepreneurial state: The redevelopment of Gunwharf Quays, Portsmouth. University of Manchester, Manchester, SPA Working Paper, 51.

Cooke, S. and Jenkins, L. (2001). Discourses of regeneration in early twentieth-century Britain: From Bedlam to the Imperial war Museum. *Area*, 33(4), 382–90.

Cooper, K. (1988). Ashburn Hall *Mental Health News*, October/November, 6–7.

Cornish, C. (1997). Behind the crumbling walls: The re-working of a former asylum's geography. *Health and Place*, 3(2), 101–10.

Cracknell, P. (2005). County Asylums: West Sussex county asylum. Available at: http://www.countyasylums.com/mentalasylums/graylingwell01.htm (accessed 2 December 2014).

Craig, T. and Timms, P. (1992). Out of the wards and onto the streets? Deinstitutionalization and homelessness in Britain. *Journal of Mental Health*, 1(3), 265–75.

Davidson, J., Bondi, L. and Smith, M. (2005). *Emotional Geographies*. London: Routledge.

Davidson, M. and Lees, L. (2005). New-build 'gentrification' and London's riverside renaissance. *Environment and Planning A*, 37(7), 1165–90.

de Certeau, M. (1984). *The Practice of Everyday Life*. Berkeley, CA: University of California Press.

Dear, M. (1976). Abandoned housing, in *Urban Policy-making and Metropolitan Dynamics: A Comparative Geographical Analysis*, edited by J. Adams. Cambridge: Ballinger.

Dear, M. (1978). Planning for mental health care: A reconsideration of public facility location theory. *International Regional Science Review*, 3(2), 93–111.

Dear, M. (1980). The public city, in *Residential Mobility and Public Policy*, edited by W. Clark and E. Moore. Beverly Hills, CA: Sage.

Dear, M., Bayne, L., Boyd, G., Callaghan, E. and Goldstein, E. (1980). *Coping in the Community: The Needs of Ex-psychiatric Patients*. Hamilton, Ontario: Mental Health.

Dear, M., Clark, G. and Clark, S. (1979). Economic cycles and mental health care policy: An examination of the macro-context for social service planing. *Social Science and Medicine*, 13(1), 43–53.

Dear, M., Fincher, R. and Currie, L. (1977). Measuring the external effects of public programs. *Environment and Planning A*, 9(2), 137–47.

Dear, M. and Wolch, J. (1987). *Landscapes of Despair: From Deinstitutionalisation to Homelessness*. Oxford: Polity Press.

Dear, M. and Taylor, S. (1982). *Not on our Street*. London: Pion.

DeSilvey, C. (2006). Observed decay: Telling stories with mutable things. *Journal of Material Culture*, 11(3), 318–38.

DeSilvey, C. (2007). Salvage memory: Constellating material histories on a hardscrabble homestead. *Cultural Geographies*, 14(3), 401–24.

DeSilvey, C. and Edensor, T. (2012). Reckoning with ruins. *Progress in Human Geography*, 37(4), 465–85.

Dodd, J. (2002). Museums and the health of the community, in *Museums, Society and Inequality*, edited by R. Sandell. London: Routledge.

Dolan, L. (1987). Reuse of state hospital property, 1970–1985. *Hospital and Community Psychiatry*, 38(4), 408–10.

Dowdall, G. (1996). *The Eclipse of the State Mental Hospital: Policy, Stigma and Organization*. Albany, NY: State University of New York Press.

Durie, M. (1999). Mental health and Maori development. *Australian and New Zealand Journal of Psychiatry*, 33(1), 5–12.

Early, D. (2003). *'The Lunatic Pauper Palace': Glenside Hospital Bristol, 1861–1994: Its birth, development and demise*. Bristol: Friends of Glenside Hospital Museum.

Edensor, T. (2005a). The ghosts of industrial ruins: Ordering and disordering memory in excessive space. *Environment and Planning D: Society and Space*, 23(6), 829–49.

Edensor, T. (2005b). *Industrial Ruins: Spaces, Aesthetics and Materiality*. London: Berg Publishers.

Edensor, T. (2005c). Waste matter-the debris of industrial ruins and the disordering of the material world. *Journal of Material Culture*, 10(3), 311–32.

English Heritage. (2006). Graylingwell Hospital, Chichester. Historic Landscape Characterisation. Available at: http://www.english-heritage.org.uk/publications/graylingwell-hlc-report/graylingwellhlcreportaug06.pdf.

Fitzpatrick, M. (2008). Stigma. *British Journal of General Practice*, 58(549), 294.

Flyvberg, B. (2006). Five misunderstandings about case study research. *Qualitative Inquiry*, 12(2), 219–45.

Foote, K. and Azaryahu, M. (2007). Toward a geography of memory: Geographical dimensions of public memory and commemoration. *Journal of Political and Military Sociology*, 35(1), 125–44.

Forest, B., Johnson, J. and Till, K. (2004). Post-totalitarian national identity: Public memory in Germany and Russia. *Social & Cultural Geography*, 5(3), 357–80.

Foucault, M. (1967). *Madness and Civilisation: A History of Madness in the Age of Reason*. London: Tavistock.

Franklin, B. (2002). Hospital – Heritage – Home: Reconstructing the nineteenth century lunatic asylum. *Housing, Theory and Society*, 19(3–4), 170–84.

Franks, A. (1998). The last chance saloon. *The Times Magazine*, 31 October, 22–32.

Galliford Try Homes. (ND). Graylingwell Park, phase one, Available at: http://gallifordtryhomes.co.uk/live/developments/graylingwell%20brochure.pdf (accessed 2 December 2014).

Garrett, B. and Hawkins, H. (2013). And now for something completely different … Thinking through explorer subject-bodies, a response to Not everyone has (the) balls: Urban exploration and the persistence of masculinist geography (Mott and Roberts, 2013). *Antipode*, 1–22.

Garrett, B. (2010). Urban explorers: Quests for myth, mystery and meaning. *Geography Compass*, 4(10), 1448–61.

Garrett, B. (2011). Assaying history: Creating temporal junctions through urban exploration. *Environment and Planning-Part D*, 29(6), 1048.

Garrett, B. (2014). Undertaking recreational trespass: Urban exploration and infiltration. *Transactions of the Institute of British Geographers*, 39(1), 1–13.

Geoghegan, H. (2010). Museum geography: Exploring museums, collections and museum practice in the UK. *Geography Compass*, 4(10), 1462–76.

Gesler, W. (1992a). Therapeutic landscapes: Medical issues in light of the new cultural geography. *Social Science and Medicine*, 34(7), 735–46.

Gesler, W. (1992b). *The Cultural Geography of Health Care*. Pittsburgh, VI: University of Pittsburgh Press.

Gesler, W. and Kearns, R. (2002). *Culture/Place/Health*. London: Routledge.

Giaccaria, P. and Minca, C. (2011). Topographies/typologies of the camp: Auschwitz as a sptial threshold. *Political Geography*, 30(1), 3–12.

Giddens, A. (1979). *Central Problems in Social Theory*. Berkeley, CA: University of California Press.

Giddens, A. (1984). *The Constitution of Society*. Cambridge: Polity Press.

Giggs, J. (1973). The distribution of schizophrenics in Nottingham. *Transactions of the Institute of British Geographers*, 59, 55–76.

Giggs, J. (1986). Mental disorders and ecological structure in Nottingham. *Social Science and Medicine*, 23(10), 945–61.

Gleeson, B. and Kearns, R. (2001). Remoralising landscapes of care. *Environment and Planning D: Society and Space*, 19(1), 61–80.

Goatcher, J. and Brunsden, V. (2011). Chernobyl and the sublime tourist. *Tourist Studies*, 11(2), 115–37.

Goffman, E. (1961). *Asylums: Essays on the Social Situation of Mental Patients*. New York: Anchor Books.

Goffman, E. (1963). *Stigma: Notes on the Management of Spoiled Identity*. New York: Simon & Schuster.

Goodwin, S. (1997). *Comparative Mental Health Policy: From Institutional to Community Care*. London: Sage.

GOSE (Government Office for the South East). (2009). *The South East Plan*. London: The Stationery Office.

Gough, P. (2004). Sites in the imagination: The Beaumont Hamel Newfoundland Memorial on the Somme. *Cultural Geographies*, 11(3), 235–58.

Gould, M. and Silverman, R. (2013). Stumbling upon history: Collective memory and the urban landscape. *Geoforum*, 78(5), 791–801.

Government of Ontario. (1998). *Mental Health Act: Revised Statutes of Ontario 1990*. Toronto: Queen's Printer.

Grant, J. and Mittelsteadt, L. (2004). Types of gated communities. *Environment and Planning B*, 31(6), 913–30.

Green Balance. (2006). *The Disposal of Heritage Assets by Public Bodies*. London: National Trust.

Gronfein, W. (1985). Psychotropic drugs and the origins of deinstitutionalisation. *Social Problems*, 32, 437–54.

Haines, H. and Abbott, M. (1985). Deinstitutionalization and social policy in New Zealand: I. Historical trends. *Community Mental Health in New Zealand*, 1(2), 44–56.

Halbwachs, M. (1992). *On Collective Memory* Chicago, IL: University of Chicago Press.

Halfacree, K. (1995). Talking about rurality: Social representations of the rural as expressed by residents of six English parishes. *Journal of Rural Studies*, 11(1), 1–20.

Hall, G. and Joseph, A. (1988). Deinstitutionalization and community mental health care in New Zealand: A non-policy analysis. *Environments*, 7, 44–56.

Hallows, N. (2011, 9 April). Devil's island or pauper place, *BMA News*.

Hamnett, C. and Whitelegg, D. (2007). Loft conversion and gentrification in London: From industrial to postindustrial land use. *Environment and Planning A*, 39(1), 106.

Hansard. (2002). *House of Commons Debates Written Answers: 10 April*.

Hartford, K., Schrecker, T., Wiktorowicz, M., Hoch, J. and Sharp, C. (2003). Four decades of mental health policy in Ontario, Canada. *Administration and Policy in Mental Health and Mental Health Services Research*, 31(1), 65–73.

Harvey, D. (1996). *Justice, Nature and the Geography of Difference*. Oxford: Blackwell.

Hay, I. (1988). *The Caring Commodity*. Auckland: Oxford University Press.

Hellier Langston. (ND). Light Villa Site. Available at: http://www.hlp.co.uk/57/95/light-villa-site?id=56 (accessed 19 October 2014).

Hoelscher, S. (2008). Angels of memory: Photography and haunting in Guatemala City. *GeoJournal*, 73(3), 195–217.

Hoelscher, S. and Alderman, D. (2004). Memory and place: Geographies of a critical relationship. *Social and Cultural Geography*, 5(3), 347–56.

Holcomb, B. (1993). Revisioning place: De- and re-constructing the image of the industrial city, in *Selling Places: The City as Cultural Capital, Past and Present*, edited by G. Kearns and C. Philo. Oxford: Pergamon Press.

Holloway, J. (2010). Legend-tripping in spooky spaces: Ghost tourism and infrastructures of enchantment. *Environment and Planning D: Society and Space*, 28(4), 618–37.

Holloway, J. and Kneale, J. (2008). Locating haunting: A ghost-hunter's guide. *Cultural Geographies*, 15(3), 297–312.

Homewood Health Centre. (1998). *Operating Plan, 1998–99.* Guelph, Ontario: Homewood Health Centre.

Homewood Health Centre. (2003a). Grounds. Available at: http://www.homewood. org/healthcentre/main.php?tID=0&sID=5&lID=2 (accessed 18 August 2005).

Homewood Health Centre. (2003b). Home. Available at: http://www.homewood. org/healthcentre/main.php?tID=0 (accessed 18 August 2005).

Homewood Health Centre. (2003c). IMAP (Integrated Mood and Anxiety Program): Ron Ellis' Story. Available at: http://www.homewood.org/healthcentre/main. php?tID=1&sID=0&lID=3 (accessed 18 August 2005).

Homewood Health Corporation. (N.D.). *Homewood: Your Behavioural Health Partner.* Guelph, Ontario: Homewood Health Centre.

Hopkins, J. (1990). West Edmonton mall: Landscape of myth and elsewhereness. *The Canadian Geographer*, 34(1), 2–17.

Hoskins, G. (2007). Materialising memory at Angel Island Immigration Station, San Francisco. *Environment and Planning A*, 39(2), 437–55.

Hudson, B. (1991). Deinstitutionalisation: What Went Wrong. *Disability & Society*, 6(1), 21–36.

Humber Institute of Technology and Advanced Learning and Assembly Hall City of Toronto. (N.D). The Lakeshore Grounds: A Community of Learning, Recreation, Creativity, Caring and Stewardship (Pamphlet).

JBEC. (1965). St James Hospital 2000 A.D. *St James' Journal* (Summer).

Johnson, N. (2008). Public memory, in *A Companion to Cultural Geography*, edited by J. Duncan, N. Johnson and R. Schein. Oxford: Blackwell Publishing, pp. 316–27.

Johnston, R. and Withers, C. (2008). Knowing our own history? Geography department archives in the UK. *Area*, 40(1), 3–11.

Johnston, R. and Sidaway, J. (1997). *Geography and Geographers: Anglo-American Geography since 1975*. Hoboken, NJ: John Wiley & Sons.

Jones, K. (1993). *Asylums and After: A Revised History of the Mental Health Services: From the Early 18th Century to the 1990s*. London and Atlantic Highlands, NJ: Athlone Press.

Jones, K. and Moon, G. (1992). Medical geography: Global perspectives. *Progress in Human Geography*, 16(4), 563–72.

Jones, K. and Moon, G. (1999). Medical geography: Taking space seriously. *Progress in Human Geography*, 17(4), 515–24.

Jonsen-Verbeke, M. (1999). Industrial heritage: A nexus for sustainable tourism development. *Tourism Geographies: An International Journal of Tourism Space, Place and Environment*, 1(1), 70–85.

Joseph, A. and Chalmers, A. (1998). Coping with rural change: Finding a place for the elderly in sustainable communities. *New Zealand Geographer*, 54(2), 28–36.

Joseph, A. and Hall, G. (1985). The locational concentration of group homes in Toronto. *The Professional Geographer*, 37(2), 143–54.

Joseph, A. and Kearns, R. (1999). Unhealthy acts: Interpreting narratives of community mental health car in Waikato, New Zealand. *Health and Social Care in the Community*, 7(1), 1–8.

Joseph, A., Kearns, R. and Moon, G. (2013). Re-imagining psychiatric asylum spaces through residential redevelopment: Strategic forgetting and selective remembrance. *Housing Studies*, 28(1), 135–53.

Joseph, A. and Boeckh, J. (1981). Locational variation in mental health care utilization dependent upon diagnosis: A Canadian example. *Social Science and Medicine*, 15D, 395–404.

Joseph, A. and Kearns, R. (1996). Deinstitutionalization meets restructuring: The closure of a psychiatric hospital in New Zealand. *Health and Place*, 2(3), 179–89.

Joseph, A., Kearns, R. and Moon, G. (2009). Recycling former psychiatric hospitals in New Zealand: Echoes of deinstitutionalisation and restructuring. *Health & Place*, 15(1), 79–87.

Joseph, A. and Moon, G. (2002). From retreat to health centre: Legislation, commercial opportunity and the repositioning of a Victorian private asylum. *Social Science and Medicine*, 55(12), 2193–200.

Kalamazoo State Hospital. (2014). Kalamazoo State Hospital. Available at: http://www.asylumprojects.org/index.php?title=Kalamazoo_ State_Hospital (accessed 18 August 2005).

Kearns, G. and Philo, C. (1993). *Selling Places: The City as Cultural Capital, Past and Present*. Oxford: Pergamon Press.

Kearns, R. (1993). Place and health: Towards a reformed medical geography. *Professional Geographer*, 45(2), 139–47.

Kearns, R. (2000). Being there: Research through observing and participating, in *Qualitative Research Methods in Human Geography*, edited by I. Hay. Melbourne: Oxford, pp. 103–21.

Kearns, R. and Barnett, J. (1999). To boldly go? Auckland's starship enterprise: Metaphors and the marketing of a children's hospital in New Zealand. *Environment and Planning D: Society and Space*, 17, 201–26.

Kearns, R., Barnett, J. and Newman, J. (2003). Placing private healthcare: Reading Ascot hospital in the landscape of contemporary Auckland. *Social Science and Medicine*, 56(11), 2303–15.

Kearns, R. and Collins, D. (2006). 'On the rocks': New Zealand's coastal bach landscape and the case of Rangitoto Island. *New Zealand Geographer*, 62(3), 227–35.

Kearns, R. and Joseph, A. (1993). Space in its place: Developing the link in medical geography. *Social Science and Medicine*, 37(6), 711–17.

Kearns, R. and Joseph, A. (1997). Restructuring health and rural communities in New Zealand. *Progress in Human Geography*, 21(1), 18–32.

Kearns, R. and Joseph, A. (2000). Contracting opportunities: Interpreting the post asylum geographies of Auckland, New Zealand. *Health & Place*, 6(3), 159–69.

Kearns, R., Joseph, A. and Moon, G. (2010). Memorialisation and remembrance: On strategic forgetting and the metamorphosis of psychiatric asylums into sites for tertiary educational provision. *Social & Cultural Geography*, 11(8), 731–49.

Kearns, R., Joseph, A. and Moon, G. (2012). Traces of the New Zealand psychiatric hospital: Unpacking the place of stigma. *New Zealand Geographer*, 68(3), 178–86.

Kearns, R., Lewis, N., McCreanor, T. and Witten, K. (2009). 'The status quo is not an option': Community impacts of school closure in South Taranaki, New Zealand. *Journal of Rural Studies*, 25(1), 131–40.

Kearns, R. and Moon, G. (2002). From medical to health geography: Novelty, place and theory after a decade of change. *Progress in Human Geography*, 26(5), 605–25.

Kearns, R. (1990). Satisfaction with community life among chronically mentally disabled persons in Auckland. *Community Mental Health in New Zealand*, 5(2), 47–63.

Kearns, R., Barnett, J. and Newman, D. (2003). Placing private health care: Reading Ascot Hospital in the landscape of contemporary Auckland. *Social Science and Medicine*, 56(11), 2303–15.

Kearns, R. and Smith, C. (1993). Housing stressors and mental health among marginalised urban populations. *Area*, 25(3), 267–78.

Kearns, R., Taylor, S. and Dear, M. (1987). Coping and satisfaction among the chronically mentally disabled. *Canadian Journal of Mental Health*, 6(2), 13–24.

Kingi, T. (2005). *Cultural Interventions and the Treatment of Māori Mental Health Consumers*. Wellington: Te Pumanawa Hauora/Research School of Public Health, Massey University.

Knapp, M., Chisholm, D., Astin, J., Lelliott, P. and Audini, B. (1997). The cost consequences of changing the hospital – community balance: The mental health residential care study. *Psychological Medicine*, 27(3), 681–92.

Landzelius, M. (2003). Commemorative dis(re)membering: Erasing heritage, spatializing disinheritance. *Environment and Planning D: Society and Space*, 21(2), 195–221.

Lawrence, J., Kearns, R., Park, J., Bryder, L. and Worth, H. (2008). Discourses of disease: Representation of tuberculosis within New Zealand newspapers 2002–2004. *Social Science and Medicine*, 66(3), 727–39.

Laws, G. and Dear, M. (1988). Coping in the community: A review of the factors influencing the lives if deinstitutionalized ex-psychiatric pations, in *Location and Stigma: Emerging Trends in the Study of Mental Health and Mental Illness*, edited by C. Smith and J. Giggs.

Le Goix, R. and Webster, C. (2008). Gated communities. *Geography Compass*, 2(4), 1189–214.

Legg, S. (2007). Reviewing geographies of memory/forgetting. *Environment and Planning A*, 39(2), 456.

Lelliott, P., Audini, B., Knapp, M. and Chisholm, D. (1996). The mental health residential care study: Classification of facilities and description of residents *British Journal of Psychiatry*, 169(2), 139–47.

Light, D. (2009). Performing Transylvania: Tourism, fantasy and play in a liminal place. *Tourist Studies*, 9(3), 240–58.

Link, B. and Phelan, J. (2001). Conceptualising stigma. *Annual Review of Sociology*, 27, 363–85.

Löwenberg-Doornbos, J. and Freedman, R. (2008). Het Dolhuys: A museum of people with mental illness and their treatment. *American Journal of Psychiatry*, 165(6), 694–694.

Lowenthal, D. (1985). *The Past is a Foreign Country*. Cambridge: Cambridge University Press.

Lowin, A., Knapp, M. and Beecham, J. (1998). Use of old long-stay hospital buildings. *Psychiatrist Bulletin*, 22(3), 129–30.

Luckhurst, R. (2002). The contemporary London Gothic and the limits of the 'spectral turn'. *Textual Practice*, 16(3), 527–46.

Lurie, S. (2005). Comparative mental health policy: Are there lessons to be learned? *International Review of Psychiatry*, 17(2), 97–101.

Macchi, S. (2003). The "Citadel of Exclusion": Regeneration processes in the area of Santa Maria della Pietà in Rome, in *Knights and Castles: Minorities and Urban Regeneration*, edited by F. Lo Piccolo and H. Thomas, 33.

Maclean, C. (2009). Wellington places. *Te Ara – the Encyclopedia of New Zealand*. Available at: http://www.teara.govt.nz/en/photograph/13498/porirua-psychiatric-hospital (accessed 2 December 2014).

Maddern, F. (2008). Spectres of migration and the ghosts of Ellis Island. *Cultural Geographies*, 15(3), 359–81.

Maddern, J. and Adey, P. (2008). Editorial: Spectro-geographies. *Cultural Geographies*, 15(3), 291–5.

Madsen, J. (1992). Place-marketing in Liverpool: A review. *International Journal of Urban and Regional Research*, 16(4), 633–40.

Mansvelt, J. (2005). *Geographies of Consumption*. London: Routledge.

Manzi, T. and Smith-Bowers, B. (2005). Gated communities as club goods: Segregation or social cohesion? *Housing Studies*, 20(2), 345–59.

Marshall, D. (2004). Making sense of remembrance. *Social and Cultural Geography*, 5(1), 37–54.

Mason, K., Bennett, H. and Ryan, A. (1988). Report of the Committee of Inquiry into procedures in certain psychiatric hospitals in relation to admission, discharge or release on leave of certain classes of patients. Wellington: Government Printer.

McCallum, D. (2008). The contingent object in psychiatry. *Philosophy, Psychiatry and Psychology*, 15(1), 69–71.

McKechnie, K. (2004). Unitec or 'Looneytec'? *Unison*, 4, 5–8.

Medlicott, J. (2001). Ashburn Hall, 1905–1947, in *Unfortunate Folk: Essays on Mental Health Treatment 1863–1992*, edited by B. Brooks and J. Thomson. Dunedin: Otago University Press, pp. 115–22.

Melling, J. and Forsythe, B. (2006). *The Politics of Madness: The State, Insanity and Society in England, 1845–1914*. London: Routledge.

Micallef, S. (2007). Mimico Asylum Walk Sunday, Spacing Toronto. Available at: http://spacing.ca/wire/2007/08/11/mimico-asylum-walk-sunday (accessed 1 July 2010).

Mohan, J. (2002). *Planning, Markets and Hospitals.* London: Routledge.

Moon, G. (1988). Is there one round here? Investigating reaction to small scale mental health hostel provision in Portsmouth, England, in *Location and Stigma: Contemporary Perspectives on Mental Health and Mental Health Care,* edited by C. Smith and J. Giggs. London: Unwin Hyman.

Moon, G. (1995). (Re)placing research on health and health care. *Health & Place*, 1(1), 1–4.

Moon, G. (2000). Risk and protection: The discourse of confinement in contemporary mental health policy. *Health and Place*, 6(3), 239–50.

Moon, G. and Brown, T. (2001). Closing Barts: Community and resistance in contemporary UK hospital policy. *Environment and Planning D*, 19(1), 43–60.

Moon, G., Kearns, R. and Joseph, A. (2006). Selling the private asylum: Therapeutic landscapes and the (re)valorization of confinement in the era of community care. *Transactions of the Institute of British Geographers*, 31(2), 131–49.

Mott, C. and Roberts, S. (2014). Not everyone has (the) balls: Urban exploration and the persistence of masculinist geography. *Antipode*, 46(1), 229–45.

Mullen, P. (2000). Forensic mental health. *The British Journal of Psychiatry*, 176(4), 307–11.

New Zealand Doctor. (2006). Māori mental health unit opens *New Zealand Doctor*, 5 April.

New Zealand Historic Places Trust (2009). Available at: http://www.historic.org.nz/aboutus/AucklandBranch/auckland_places2visit_suburbs.html#24 (accessed 1/9/09).

NHS Commissioning Board. (2013). NHS standard contract for medium and low secure mental health services (adults). Schedule 2 – the services – service specifications.

NHS Property Services, S. N. T. a. C. H. (2014). Portsmouth Community Care Estate Review: Report for The Health Overview Scrutiny Panel.

Nora, P. (1989). Between memory and history: Les lieux de mémoire. *Representations*, 26, 7–25.

Ontario Shores. (2011). A brief history. Available at: http://www.ontarioshores.ca/UserFiles/Servers/Server_6/File/PDFs/Detailed_History.pdf (accessed 29 July 2014).

Opit, S. and Kearns, R. (2014). Selling a natural community: Exploring the role of representations in promoting new urban development. *New Zealand Geographer*, 70(2), 91–102.

Paine, C. (1997). Landscapes for mental health: Design and adaptation of early psychiatric hospitals. *Proceedings of the American Society of Landscape Architects*, 211–15.

Parr, H. (2000). Interpreting the 'hidden social geographies' of mental health: Ethnographies of inclusion and exclusion in semi-institutional places. *Health & Place*, 6(3), 225–37.

Parr, H. and Philo, C. (1996). 'A forbidding fortress of locks, bars and padded cells': Locational history of mental health care in Nottingham. *Historical Geography Research Series – Institute of British Geographers*, 32, 98.

Payne, C. and Sacks, O. (2009). *Asylum: Inside the Closed World of State Mental Hospitals*. Boston, MA: MIT Press.

Phelan, J. (2005). Geneticization of deviant behavior and consequences for stigma: The case of mental illness. *Journal of Health and Social Behaviour*, 46(4), 307–22.

Phillips, D. and Vincent, J. (1986). Petit bourgeois care: Private and residential care for the elderly. *Policy and Politics*, 14(2), 189–208.

Philo, C. (1987a). "Fit localities for an asylum": The historical geography of the nineteenth-century "mad-business" in England as viewed through the pages of the Asylum Journal. *Journal of Historical Geography*, 13(4), 398–415.

Philo, C. (1987b). 'Not at our seaside': Community opposition to a nineteenth century branch asylum (England). *Area*, 19(4), 297–302.

Philo, C. (1997). Across the water: Reviewing geographical studies of asylums and other mental health facilities. *Health and Place*, 3(2), 79–89.

Philo, C. (2000). Post-asylum geographies: An introduction. *Health and Place*, 6(3), 135–6.

Philo, C. (2004). *A Geographical History of Institutional Provision for the Insane from Medieval Times to the 1860s in England and Wales: The Space Reserved for Insanity*. Lampeter: Edwin Mellen Press.

Pile, S. (2009). Emotions and affect in recent human geography. *Transactions of the Institute of British Geographers*, 35(1), 5–20.

Pilgrim, D. and Rogers, A. (2001). *Mental Health Policy in Britain*. Basingstoke: Palgrave.

Pinder, D. (2005). Arts of urban exploration. *Cultural Geographies*, 12(4), 383–411.

Pinfold, V. (2000). "Building up safe havens … all around the world": Users' experiences of living in the community with mental health problems. *Health & Place*, 6(3), 201–112.

Porteous, J.D. (1985). Smellscape. *Progress in Human Geography*, 9(3), 356–78.

Porteous, J.D (1990). *Landscapes of the Mind: Worlds of Sense and Metaphor*. Toronto: University of Toronto Press.

Portsmouth City Council. (2006). Portsmouth City Local Plan. Portsmouth: Porsmouth City Council.

Portsmouth City Council. (2013). Site Allocations. Portsmouth: Portsmouth City Council.

Portsmouth City Council. (2014). Consultation with service users, families and carers in respect of the proposal to close the Lowry Centre, Solent NHS Trust, and support people to access appropriate services in alternatie settings. Portsmouth: Portsmouth City Council.

Powell, J. (1961). The Water Tower Speech *Annual Conference: National Association for Mental Health.* Brighton.

Pred, A. (1984). Place as historically contingent process: Structuration and the time-geography of becoming places. *Annals of the Association of American Geographers*, 74(2), 279–97.

Prins, S. (2011). Does transinstitutionalization explain the overrepresentation of people with serious mental illness in the criminal justice system? *Community Mental Health Journal*, 47(6), 716–22.

Priory Group. (2005a). Overview. Available at: http://www.prioryhealthcare. co.uk/About-us/Overview (accessed 18 August 2005).

Priory Group. (2005b). Priory Grange Heathfield. Available at: http://www. prioryhealthcare.co.uk/Find-a-centre/Facilities/Priory-Grange-Heathfield (accessed 18 August 2005).

Priory Group. (2005c). Priory Hospital Altrincham. Available at: http://www. prioryhealthcare.co.uk/Find-a-centre/Facilities/Priory-Hospital-Altrincham (accessed 18 August 2005).

Priory Group. (2005d). Priory Hospital Bristol. Available at: http://www. prioryhealthcare.co.uk/Find-a-centre/Facilities/Priory-Hospital-Bristol (accessed 18 August 2005).

Priory Group. (2005e). Priory Grange Hemel Hempstead. Available at: http:// www.prioryhealthcare.co.uk/Find-a-centre/Facilities/Priory-Grange-Hemel-Hempstead (accessed 18 August 2005).

Priory Group. (2005f). The Priory Group's History. 2005. Available at: http://www. prioryhealthcare.co.uk/About-us/The-Priory-Groups-History (accessed 18 August).

Priory Group. (2005g). Priory Hospital Hayes Grove. Available at: http://www. prioryhealthcare.co.uk/Find-a-centre/Facilities/Priory-Hospital-Hayes-Grove (accessed 18 August 2005).

Priory Group. (2005h). Priory Hospital Glasgow. Available at: http://www. prioryhealthcare.co.uk/Find-a-centre/Facilities/Priory-Hospital-Glasgow (accessed 18 August 2005).

Priory Ticehurst House. (2005). Statement of purpose. Wadshurst: The Priory Ticehurst House.

Radio New Zealand. (2006). Waitemata District Health Board to open a secure intellectual disability unit at the Mason Clinic. Wellington, New Zealand: Radio New Zealand.

Ramsey, C., Carpenter, P. and Beveridge, A. (2008). Untitled pictures (date unknown) by Denis Reed (1917–1979) – psychiatry in pictures. *The British Journal of Psychiatry*, 193(4), 278.

Relph, E. (1976). *Place and Placelessness*. London: Pion.

Rodaway, P. (1994). *Sensuous Geographies: Body, Sense and Place*. London: Routledge.

Roehampton Priory Hospital. (2005). Introducing the Priory Hospital. Available at: http://www.priory-hospital.co.uk/htm/priory.htm (accessed 18 August 2005).

Rose, G. (2001). *Visual Methodologies: An Introduction to Interpreting Visual Materials*. London: Sage.

Rose, M. (2006). Gathering 'dreams of presence': A project for the cultural landscape. *Environment and Planning D: Society and Space*, 24(4), 537–54.

Salmond Architects. (1993). *Former Carrington Psychiatric Hospital, Avondale: A Conservation Plan* Auckland: Salmons Reed.

Sandell, R. (2003). *Museums, Society, Inequality*. London: Routledge.

Sandell, R., Dodd, J. and Garland-Thomson, R. (2013). *Re-presenting Disability: Activism and Agency in the Museum*. London: Routledge.

SAVE. (1995). *Mind over Matter: A Study of the Country's Threatened Mental Asylums*. London: SAVE Britain's Heritage.

Schama, S. (1995). *Landscape and Memory*. London: Harper Collins.

Scull, A. (1984). *Decarceration: Community Treatment and the Deviant: A Radical View*. New Brunswick, NJ: Rutgers University Press.

Sealy, P. and Whitehead, P. (2004). Forty years of deinstitutionalization of psychiatric services in Canada: An empirical assessment. *Canadian Journal of Psychiatry*, 49(4), 249–57.

Seligman, M. (1972). Learned helplessness. *Annual Review of Medicine*, 23(1), 407–12.

Sennett, R. (1992). *The Uses of Disorder: Personal Identity and City Life*. New York: W.W. Norton.

Silverman, D. (ed.). (2010). *Qualitative Research*. London: Sage.

Simmel, G. (1995). The metropolis and mental life, in *Metropolis: Centre and Symbol of our Times*, edited by P. Kasnitz. London: Macmillan, pp. 30–45.

Sinclair, I. (2002). *London Orbital: A Walk around the M25*. Oxford: Granta Books.

Sinclair, I. (2006). *Edge of the Orison: In the Traces of John Clare's' Journey Out of Essex'*. London: Penguin.

Skegg, P. and Cox, B. (1991). Impact of psychiatry services on prison suicide. *The Lancet*, 338(8780), 1436–8.

Smith, C. and Giggs, J. (1988). *Location and Stigma: Contemporary Perspectives on Mental Health and Mental Health Care*. Boston, MA: Allen & Unwin.

Smith, C. and Hanham, R. (1981). Any place but here! Mental health facilities as noxious neighbours. *The Professional Geographer*, 33(3), 326–34.

Smyth, F. (2005). Medical geography: Therapeutic spaces, places and networks. *Progress in Human Geography*, 29(4), 488–95.

St Joseph's Healthcare Hamilton. (2014). West 5th Campus. Available at: http://www.stjoes.ca/about/our-locations/west-5th-campus (accessed 19 October 2014).

Stearn, M., Ryall, M., Drago, M., Kidd, J. and Cairns, M. (2014). *Challenging History in the Museums: International Perspectives*. Farnham: Ashgate Publishing.

Stenning, A., Charlesworth, A., Guzik, R. and Paszkowski, M. (2008). A tale of two institutions: Shaping Oswiecim-Auschwitz. *Geoforum*, 39(1), 401–13.

Stephenson, J., Abbott, M. and Ruru, J. (eds). (2010). *Beyond the Scene: Landscape and Identity in Aotearoa New Zealand*. Dunedin: Otago University Press.

Stuff. (2010). New Zealand's spookiest stories. Available at: http://www.stuff. co.nz/life-style/4285667/New-Zealands-spookiest-stories   (accessed   2 December 2014).

Summerby-Murray, R. (2002). Interpreting deindustrialised landscapes of Atlantic Canada: Memory and industrial heritage in Sackville, New Brunswick. *Canadian Geographer*, 46(1), 48–62.

Takahashi, L. and Dear, M. (1997). The changing dynamics of community opposition to human service facilities. *Journal of the American Planning Association*, 36(1), 79–93.

Tatham, R. (1983). The Homewood: 100 years of service, in The Homewood Centennial Committee (ed.), *The Homewood Sanitarium: 100 Years of Service 1883–1983*. Guelph, Ontario: The Homewood Sanitarium.

Taylor Hazell Architects Ltd. (2007). Humber College Lakeshore Campus. Available at: http://www.taylorhazell.com/portfolio/humber.swf (accessed 1 July 2010).

Taylor, J. (1991). *Hospital and Asylum Architecture in England, 1840–1917: Building for Health care*. London and New York: Mansell.

Tepou. (N.D.). Tepou o te whakaaro nui. Available at: www.tepou.co.nz (accessed 2 December 2014).

Thinking Ahead. (2014a). Greefingers – Shaw Trust. Available at: http://www. thinkingahead.org.uk/services/others/greenfingers.htm (accessed 19 October 2014).

Thinking Ahead. (2014b). A history of mental health services in Portsmouth. Available at: http://www.thinkingahead.org.uk/history/index.htm (accessed 19 October 2014).

Till, K. (1999). Staging the past: Landscape designs, cultural identity and Erinnerungspolitik at Berlin's Neue Wache. *Cultural Geographies*, 6(3), 251–83.

Till, K. (2005). *The New Berlin: Memory, Politics, Place*. Minneapolis, MN: University of Minnesota Press.

Todd, H. (2000). The drive for quality: Talking sense. Available at: http://www. sense.org.uk/publications/magazine/tsarticles/2000/qualdrive.htm   (accessed 18 August 2005.

Toronto Culture. (N.D.). Public Art on the Lakeshore Grounds (Pamphlet).

Trieman, N. and Leff, T. (1998). Closing psychiatric hospitals – some lessons from the TAPS project, in *Residential Care versus Community Care: The Role of Institutions in Welfare Provision*, edited by R. Jack. London: Macmillan, pp. 41–52.

Tunbridge, J. (2004). The Churchill-Roosevelt bases of 1940: The question of heritage in their adaptive reuse. *International Journal of Heritage Studies,* 10(3), 229–51.

Unitec. (1994). Rededication of former Carrington Hospital Building. *Korero (Unitec Staff magazine)* 26 September.

Unitec. (N.D.a). Mount Albert Campus. Available at: http://www.unitec. ac.nz/?862D5A54-680E-4BF8-BD74-4FBFB1BCC842 (accessed 1 July 2010).

Unitec. (N.D.b). History. Available at: http://www.unitec.ac.nz/?B8F6A6BE-4E9C-49BD-8750-AF508A580E28 (accessed 1 July 2010).

Unitec. (N.D.c). A building known as Carrington.

Unitec. (N.D.d). Guidelines for the adaptation of the former Carrington Psychiatric Hospital for the Departments of Architecture, Planning and Design.

University of Leicester. (2014). The Leicestershire origins. Available at: http://www. leics.gov.uk/index/leisure_tourism/local_history/recordoffice/recordoffice_ exhibitions/record_office_towers_hospital/record_office_towers_hospital-3. htm (accessed 19 October 2014).

University of Queensland (2014). Ipswich Campus: progression of an institution. Available at: https://www.library.uq.edu.au/ipswich/uqihistory (accessed 1 February 2015).

Urry, J. (1990). *The Tourist Gaze: Leisure and Travel in Contemporary Societies.* Thousand Oaks, CA: Sage.

Van Hoven, B. (2006). Prison heritage: Museums and memorials. *Nederlandse Geografische Studies,* 39–47.

Vayda, E. and Deber, R. (1992). The Canadian health-case system: A developmental overview, in *Canadian Health and the State,* edited by D. Naylor. Montreal: McGill-Queen's University Press.

Viejo-Rose, D. (2011). Memorial functions: Intent, impact and the right to remember. *Memory Studies,* 4(4), 465–80.

Waitemata District Health Board. (N.D.). Auckland Regional Forensic Psychiatry Services. Available at: http://www.waitematadhb.govt.nz/PatientsVisitors/ TheMasonClinic.aspx.

Waitt, G. and McGuirk, P. (1997). Selling waterfront heritage: A critique of Millers Point, Sydney. *Tijdschrift voor Economische en Sociale Geografie,* 88(4), 342–52.

Watson, S. and Wells, K. (2005). Spaces of nostalgia: The hollowing out of a London market. *Social and Cultural Geography,* 6(1), 17–30.

Weiner, D. (2004). The erasure of history: From Victorian asylum to 'Princess Park Manor', in *Architecture as Experience: Radical Change in Spatial Practice,* edited by D. Arnold and A. Ballantyne. London: Routledge.

Wellington-Dufferin District Health Council. (1996). *An Assessment of Mental Health Needs and Resources in Wellington – Dufferin.* Guelph, Ontario: The Wellington Dufferin District Health Council.

Williams, A. (1999). *Therapeutic Landscapes: The Dynamic between Place and Wellness.* Lanham, MD: University Press of America.

Williams, D. (2008). *Abandoned Insane Asylums*. New York: Bearport Publishing Company.

Williams, J. and Lutterbach, E. (1976). The changing boundaries of psychiatry in Canada. *Social Science and Medicine*, 10(1), 15–22.

Wilson, K. (2003). Therapeutic landscapes and First Nations peoples: An exploration of culture, health and place. *Health & Place*, 9(2), 83–93.

Withers, C. (2006). History and philosophy of geography, 2003–2004: Geography's modern histories? International dimensions, national stories, personal accounts. *Progress in Human Geography*, 30(1), 79–86.

Wolch, J. and Philo, C. (2000). From distributions of deviance to definitions of difference: Past and future mental health geographies. *Health and Place*, 6(3), 137–57.

Woodburn Priory Hosptial. (2004). Statement of purpose. Birmingham: Woodbourne Priory Hospital.

Wylie, J. (2007). The spectral geographies of W.G. Sebald. *Cultural Geographies*, 14(2), 171–88.

Wylie, J. (2009). Landscape, absence and the geographies of love. *Transactions of the Institute of British Geographers*, 34(3), 275–89.

Yanni, C. (2007). *The Architecture of Madness: Insane Asylums in the United States*. Minneapolis, MN: University of Minnesota Press.

Yin, R. (1989). *Case Study Research: Design and Method*. London: Sage.

Index

References to tables are shown in **bold** and those for figures are in *italics*.

3 20